Lecture Notes in Networks and Systems

Volume 592

The series "Lecture Notes in Networks and Systems" publishes the latest developments in Networks and Systems—quickly, informally and with high quality. Original research reported in proceedings and post-proceedings represents the core of LNNS.

Volumes published in LNNS embrace all aspects and subfields of, as well as new challenges in, Networks and Systems.

The series contains proceedings and edited volumes in systems and networks, spanning the areas of Cyber-Physical Systems, Autonomous Systems, Sensor Networks, Control Systems, Energy Systems, Automotive Systems, Biological Systems, Vehicular Networking and Connected Vehicles, Aerospace Systems, Automation, Manufacturing, Smart Grids, Nonlinear Systems, Power Systems, Robotics, Social Systems, Economic Systems and other. Of particular value to both the contributors and the readership are the short publication timeframe and the world-wide distribution and exposure which enable both a wide and rapid dissemination of research output.

The series covers the theory, applications, and perspectives on the state of the art and future developments relevant to systems and networks, decision making, control, complex processes and related areas, as embedded in the fields of interdisciplinary and applied sciences, engineering, computer science, physics, economics, social, and life sciences, as well as the paradigms and methodologies behind them.

Indexed by SCOPUS, INSPEC, WTI Frankfurt eG, zbMATH, SCImago.

All books published in the series are submitted for consideration in Web of Science.

For proposals from Asia please contact Aninda Bose (aninda.bose@springer.com).

Damir Blažević · Naida Ademović ·
Tomislav Barić · Josip Cumin · Eleonora Desnica
Editors

31st International Conference on Organization and Technology of Maintenance (OTO 2022)

Springer

Editors
Damir Blažević
Faculty of Electrical Engineering, Computer Science and Information Technology
Josip Juraj Strossmayer University of Osijek
Osijek, Croatia

Tomislav Barić
Faculty of Electrical Engineering, Computer Science and Information Technology
Josip Juraj Strossmayer University of Osijek
Osijek, Croatia

Eleonora Desnica
Technical Faculty "Mihajlo Pupin"
University of Novi Sad
Zrenjanin, Serbia

Naida Ademović
Faculty of Civil Engineering
University of Sarajevo
Sarajevo, Bosnia and Herzegovina

Josip Cumin
Faculty of Mechanical Engineering
University of Slavonski Brod
Slavonski Brod, Croatia

ISSN 2367-3370 ISSN 2367-3389 (electronic)
Lecture Notes in Networks and Systems
ISBN 978-3-031-21428-8 ISBN 978-3-031-21429-5 (eBook)
https://doi.org/10.1007/978-3-031-21429-5

This Springer imprint is published by the registered company Springer Nature Switzerland AG
The registered company address is: Gewerbestrasse 11, 6330 Cham, Switzerland

Conference Organizers and Board Members OTO 2022

International Program Committee

Scientific Committee

izv. prof. dr. sc. Naida Ademović, BiH
dr. sc. José-Lázaro Amaro-Mellado, Spain
dr. sc. Ivan Ambroš, Croatia
izv. prof. dr. sc. Samir Avdaković, BiH
doc. dr. sc. Josip Balen, Croatia
izv. prof. dr. sc. Tomislav Barić, Croatia
prof. dr. sc. Đuro Banaj, Croatia
izv. prof. dr. sc. Marinko Barukčić, Croatia
izv. prof. dr. sc. Huseyin Bilgin, Albania
izv. prof. dr. sc. Damir Blažević, Croatia
izv. prof. dr. sc. Mirjana Bošnjak-Klečina, Croatia
izv. prof. dr. sc. Borko Bulajić, Serbia
prof. dr. sc. Dominika Crnjac-Milić, Croatia
dr. sc. Slawomir Czarnecki, Poland
doc. dr. sc. Josip Cumin, Croatia
prof. dr. sc. Sanja Dimter, Croatia
prof. dr. sc. Zlata Dolaček-Alduk, Croatia
doc. dr. sc. Tihomir Dokšanović, Croatia
prof. dr. sc. Snežana Đurđić, Serbia
prof. dr. sc. Mohamed Elchalakani, Australia
izv. prof. dr. sc. Krešimir Fekete, Croatia
izv. prof. dr. sc. Irena Galić, Croatia
izv. prof. dr. sc. Ranko Gantner, Croatia
doc. dr. sc. Mario Galić, Croatia
doc. dr. sc. Goran Gazić, Croatia
izv. prof. dr. sc. Hrvoje Glavaš, Croatia
dr. sc. Ivan Grgić, Croatia
izv. prof. dr. sc. Krešimir Grgić, Croatia
dr. sc. Ehsan Harirchian, Germany
izv. prof. dr. sc. Bassam A. Tayeh, Palestine
dr. sc. Ivana Hartmann Tolić, Croatia
prof. dr. sc. Mostafa Fahmi Hassanein, Egypt
prof. dr. sc. Željko Hocenski, Croatia
izv. prof. dr. sc. Ercan Işık, Turkey
izv. prof. dr. sc. Irena Ištoka Otković, Croatia
izv. prof. dr. sc. Aleksandar Jurić, Croatia
izv. prof. dr. sc. Josip Job, Croatia
doc. dr. sc. Željka Jurković, Croatia
izv. prof. dr. sc. Tanja Kalman Šipoš, Croatia
izv. prof. dr. sc. Tomislav Keser—President, Croatia
izv. prof. dr. sc. Zvonimir Klaić, Croatia
doc. dr. sc. Goran Knežević, Croatia

Organizing Committee

mr. sc. Tatjana Mijušković-Svetinović—President
mr. sc. Franjo Ambroš
dr. sc. Milan Ivanović
Ružica Kljajić, mag.ing.el.
prof. dr. sc. Zlatko Lacković
doc. dr. sc. Kristina Mastanjević
Ana Šokčević, prof.

Editorial Board

izv. prof. dr. sc. Damir Blažević—President
izv. prof. dr. sc. Naida Ademović
izv. prof. dr. sc. Tomislav Barić
doc. dr. sc. Josip Cumin
prof. dr. sc. Eleonora Desnica

Technical and IT support

Davor Begić, bacc.ing.
Dario Došen, mag.ing.
Ružica Kljajić, mag.ing.
Mario Miloloža, mag.ing.
Igor Sušenka, dipl.ing.

Contents

Noise Removal from Images by Applying Deep Neural Networks 1
Dijana Ivezic and Caslav Livada

Conceptual Design of Smart Furniture Objects Cluster 18
Domagoj Bilandžija, Damir Blažević, Tomislav Keser, and Andrej Živković

Organization of a Digital Voting System Based on Blockchain Technology for the Faculty Council 25
Mirko Köhler, Miljenko Švarcmajer, Ivica Lukić, and Tomislav Stipanić

Mobile Application for Keeping Records of Class Attendance 37
Petar Nenadić, Damir Blažević, and Tomislav Keser

Comparison Functionalities of HTTP and MQTT Protocols 45
Tomislav Kušević, Damir Blažević, and Tomislav Keser

Influence of Wind Power Plant on P-V and Q-V Curves in Transmission System 57
Filip Pisonic, Kresimir Fekete, Ruzica Kljajic, and Predrag Maric

New Trends in Power Transformer Surveillance and Diagnostics in the Function of Power System Maintenance 69
Igor Provči, Robert Noskov, and Ivica Petrović

Requirements for High-Quality Thermal Inspection of the Transmission Lines 84
Igor Lukačević and Domagoj Lagator

The Influence of Preventive Maintenance of Batteries on Increasing the Security of the Thermal Power System 96
Duška Jarčević, Ljiljana Radovanović, Jasmina Pekez, Borivoj Novaković, and Hrvoje Glavaš

Reverse Engineering and 3D Printing of the Spinning Reel: From Maintenance to the New Product Design ... 106
Josip Vukašinović, Ivan Grgić, Jure Marijić, and Mirko Karakašić

Development of V-Twin Motorcycle Valve Stem Seal Non-invasive Removal Tool ... 118
Josip Cumin, Daniel Novoselović, Dejan Marić, and Tomislav Šolić

Selection, Dimensioning and Maintenance of Roller Bearings ... 133
Eleonora Desnica, Aleksandar Ašonja, Ljiljana Radovanović, Ivan Palinkaš, and Imre Kiss

Methode for Quick Determination of the Reliability Level of Agricultural PTO Shafts ... 143
Aleksandar Ašonja, Eleonora Desnica, Alexander Pastukhov, Yury Kuznetsov, and Igor Kravchenko

Regeneration of Mould Inserts Using the Laser Welding Method ... 150
Waldemar Matysiak, Miroslav Duspara, Grzegorz Stawiński, Josip Cumin, Karol Malutkiewicz, Antun Stoić, and Jakub Hajkowski

Estimation of Welding Costs During the Production of the Lower Pedestal Central Ring of the Hydraulic Excavator ... 160
Sara Radojicic, Pejo Konjatic, Marko Katinic, Josip Kacmarcik, and Andrijana Milinović

Automatic Control of Recirculation System for Respiratory Control ... 169
Miloš Jovanović, Djordje Dihovicni, and Muharrem Hilmi Aksoy

Draught Beer—Maintaining the Quality Through Hygienic Measures ... 176
Kristina Habschied, Snježana Keleković, Ana Domaćinović, Tanja Mađarević Pavetić, and Krešimir Mastanjević

Innovation Management as an Important Segment of the Maintenance and Business Development in IT Companies ... 186
Dominika Crnjac Milić

Improvement and Maintenance of Communication Tools in Micro-enterprises ... 198
Katarina Štavlić, Krešimir Lacković, and Igor Štavlić

Author Index ... 213

About the Editors

Damir Blažević (1977) was born in Đakovo, Croatia. He received his B.E. degree in electrical engineering in 2001, degree master of sciences in electrical engineering, was received in 2006, and Ph. D. degree in electrical engineering, telecommunications and informatics was received in 2012, all at the Faculty of Electrical Engineering in Osijek, Croatia. He is a member of the Croatian Chamber of Architects and Engineers in Construction since 2007 and a member of several Technical Committees of the Croatian Standards Institute and participates in teaching undergraduate and graduate level at Faculty of Electrical Engineering, Computer Science and Information Technology Osijek. His research interests include computer networks, expert and intelligent systems, maintenance technology, and renewable energy sources. Currently, he is an associate professor at the Josip Juraj Strossmayer University of Osijek and the chair of the Department of Computer Engineering at Faculty of Electrical Engineering, Computer Science and Information Technology Osijek.

Naida Ademović (1973) was born in Sarajevo, Bosnia, and Herzegovina, and is an associate professor at the University of Sarajevo, Faculty of Civil Engineering. She gained her degrees at the University of Sarajevo, Faculty of Civil Engineering, Bosnia and Herzegovina (Bachelor of Sciences and Ph.D.), Rühr Universität, Fakultät für Bauingenieurwesen, in Bochum (Germany) (Master of Science); University of Padova (Italy) and University of Minho, Guimarães (Portugal) (Advanced Master of Science). Her scientific and professional activity covers the field of earthquake engineering, concrete and masonry structures, and bridges, finite element modeling, and resilience of structures. She worked on numerous professional and scientific local and international projects. She is a member of scientific committees of international journals, and she was involved in organizing several conferences. As well, she is an active reviewer for several international journals. She worked on projects regarding Integrating Seismic Risk Considerations into Energy Efficiency Investments in the Western Balkans and Improving Sarajevo's Resilience through Urban Regeneration. She has participated in several scientific projects and has been engaged in research in the field of concrete and masonry structures, earthquake engineering, and earthquake risk assessment. She is an author/co-author of five books. She participated in the establishment of an interdisciplinary study program Protection against Natural Disasters at the Center for Interdisciplinary Studies, University of Sarajevo, where she is an associate professor. She is a member of the Ph.D. study course at the University of Padova (Italy). She has published more than 100 scientific and professional papers.

Tomislav Barić (1975) was born in Našice, Croatia. He obtained B.Sc. (2000), Master of Science Degree (M.Sc.) (2004), and Ph.D. (2007) from Faculty of Electrical Engineering Osijek, the University of Osijek, Republic of Croatia. His research interests are power systems engineering, measurements, numerical methods in electromagnetic, supercapacitor modeling, and supercapacitor and battery power management systems. His employment experience included the scientific research for HEP group (Croatian national energy company). Currently, he is a full professor at the Josip Juraj Strossmayer University of Osijek and the chair of Electric Machines and Power Electronics, Faculty of Electrical Engineering, Computer Science, and Information Technology Osijek (former Faculty of Electrical Engineering Osijek).

Josip Cumin (1984) was born in Slavonski Brod, Croatia. He received his Ph.D. degree in 2014 in Mechanical Engineering from the Josip Juraj Strossmayer University of Osijek, Mechanical Engineering Faculty in Slavonski Brod. His fields of interest are metal forming technologies, tool construction, finite element method analysis, and CAD modeling. Josip Cumin has been employed at the Mechanical Engineering Faculty in Slavonski Brod, Josip Juraj Strossmayer University of Osijek since 2008. In 2008, he was employed as a research assistant, and in 2014, he was appointed an assistant professor. Currently, he is an assistant professor at the University of Slavonski Brod, Mechanical Engineering Faculty. The list of his scientific and research papers is available at the link: https://www.bib.irb.hr/pregled/profil/23891

Eleonora Desnica (1971) was born in Zrenjanin, Serbia. She graduated at the University of Novi Sad, Faculty of technical sciences in Novi Sad, Mechanical Engineering. She achieved M.Sc. (in 2004) and Ph.D. degree in technical science in 2010 at Technical faculty "Mihajlo Pupin" Zrenjanin, University of Novi Sad. Since 1998, she has been employed at Technical Faculty, "Mihajlo Pupin" in Zrenjanin. She is a member of Department of Mechanical Engineering and became a full professor in 2021. Her fields of work are as follows: computer supported technologies in mechanical engineering and their application in education, CAD/CAM, machine design, methodology of engineering design, thermography in industrial practice, mechanics and machine mechanisms, and engineering graphic communications. In her professional and scientific work, she has published as the author or the co-author over 100 papers in national and international journals and reports at scientific professional meetings. She is a member of scientific committees of national and international journals and organizing several conferences. She works as a researcher on two projects—one domestic and one international.

Noise Removal from Images by Applying Deep Neural Networks

Dijana Ivezic and Caslav Livada(✉)

Faculty of Electrical Engineering, Computing and Information, Osijek, Croatia
dijana.ivezic@etfos.hr, caslav.livada@ferit.hr

Abstract. This article addresses the problem of removing noise from images using neural networks. The type of noise that is removed in this article is salt and pepper noise. This noise represents jumps in intensity in the image, usually caused by data transmission errors. To quantify the success of the noise removal model, PSNR metrics are used. The paper provides a theoretical basis for the task in the form of a description of the neural network and its components, as well as its operation. It also describes deep learning and the differences and advantages of each type. The neural network used in this paper is based on the U-Net architecture. Using the Python programming language and the Keras application user interface, a model was developed to remove added white Gaussian noise and salt and pepper noise of varying intensity from images in the CIFAR-10 and MNIST image databases. The model is applied to the author's image that is not included in the tested databases. The analysis is presented in the PSNR presented for different noise intensities and examples of images with and without noise.

Keywords: Neural networks · Noise removal · Keras

1 Introduction

Image noise is any random change in image brightness that degrades image quality or makes it impossible to read the desired data. It is an unwanted signal added to a pure image signal. The image may be so noisy that details or even the entire image are no longer discernible. Some noise reduction solutions only blur the image and those important details are no longer visible; we lose the important information that led to the capture of the image. Examples of such solutions are the median filter or Gaussian smoothing. Computers can be taught to remove noise in a different way by applying Deep Learning. Deep Learning is a branch of artificial intelligence science that allows us to understand language by computer, recognize images, make decisions based on data, etc. What would be easy for a human to do is very difficult for a computer to do.

Deep Learning gets its name from the fact that the layers of the neural network used go deep, into multiple layers that we can understand by imagining how each layer is responsible for recognizing certain features of images. This work is about using Deep Learning to remove noise from images. When a noisy image is imagined, the parts of the image may not be visible at all under the noise. These parts need to be predicted

D. Blažević et al. (Eds.): OTO 2022, LNNS 592, pp. 1–17, 2023.
https://doi.org/10.1007/978-3-031-21429-5_1

to eventually get a clean and complete image and are predicted by using the rest of the image, blind or non-blind. In non-blind image filling, the algorithm does not know exactly where the irregularities are, which requires human work. In blind image filling, the algorithm automatically detects the noise to be removed, which makes the system more complex, but more interesting from the point of view of automating noise removal, which is very important for a large number of images. Deep learning learns system parameters from data without human intervention. This work is based on two types of noise: additive white Gaussian noise and salt and pepper noise. Peak signal to noise ratio (PSNR) is used to quantify the success of the denoising model. The higher the PSNR value, the better the quality of the reconstructed image.

2 Related Work

2.1 Real Image Denoising with Feature Attention

According to Anwar and Barnes [1], neural network models often work well on data generated during training, i.e. data obtained by artificially adding noise to a clear image, and are less successful on actual data, which is why the networks are trained. The practicality of such networks is limited, which they exploit in their work by presenting the RIDNet network for blind noise removal. They present a model that ensures top results using a modular network where increasing the number of modules helps improve performance, unlike most previous solutions.

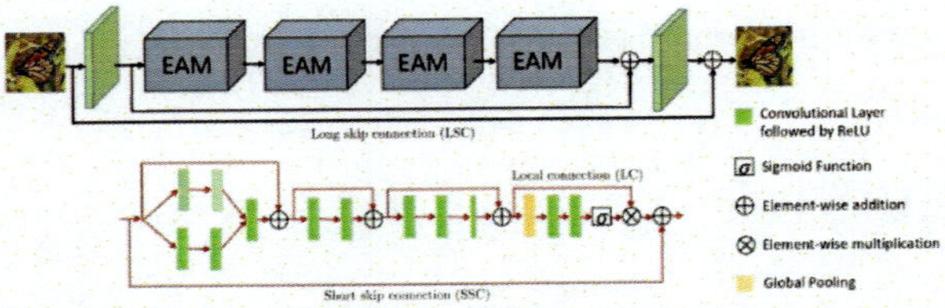

Fig. 1. Proposed architecture [1].

The upper part of Fig. 1 shows the essence of the work of the model. The modules for property extraction and reconstruction are marked in green; at the beginning and at the end. They consist of only one convolutional layer. Following the arrows, we see that after the feature extraction come the EAM ("enhancement attention modules") blocks, which are the main part of a separate feature learning module and whose task is to enhance attention to the features of the whole model. The EAM blocks are shown one by one and there are four of them in the proposed network, it is possible to add more to improve the performance. So the network can be extended to create a very deep network if needed [1].

A comparison of the presented model with other models (the presented model is shown under "Ours", last picture) is shown in Fig. 2. The first image (shown under "Noisy") is an image with noise. The image is taken from the RNI15 database.

The success of the proposed network is clearly visible, especially in the details shown in the figure. In their paper [1], more detailed comparisons are presented as well as the PSNR value versus other solutions.

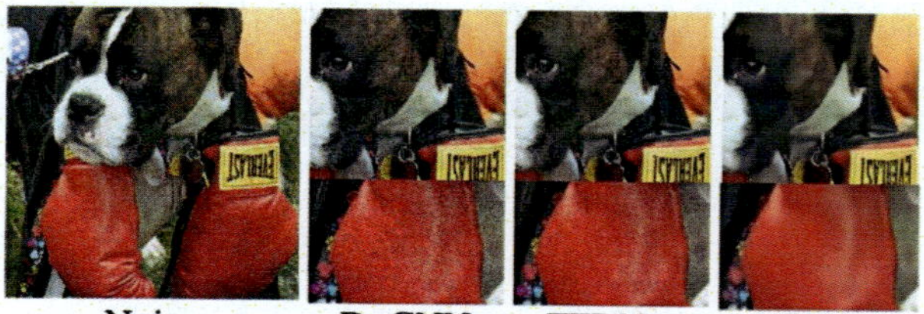

Fig. 2. Comparison with other architectures [1].

2.2 A Physics-Based Noise Formation Model for Extreme Low-Light Raw Denoising

Learned algorithms for removing image noise have difficulty using actual images from cameras or other sources because they are trained with training images that do not follow real noise well. The problem that Vei et al. [2] are addressing is modeling noise that is very similar to noise in real images, mainly noise caused by electronics in digital devices and noise that occurs in low light. It is clear that collecting real data is a tedious and expensive task, and that the quality of the modeled data greatly affects the usability of systems trained on real images using this data. By studying the physical processes involved in the generation of real images, they have developed a better model for the added noise in images based on CMOS photosensors that describes the real noise well. The obtained data show that the existing models for noise removal from images trained on the obtained more realistic data are more successful. The researchers [2] also introduce the adjustment of the parameters of their noise generation to describe as accurately as possible the noise generated on the selected vendor's camera.

Figure 3 shows four different stages in which different types of noise are inevitably added to the photons from nature in the digital camera. In the blue part, the recording noise is added to the photons entering the camera lens because of the nature of the light particles. In the orange part, which refers to electrons, the signal from the previous stage is multiplied by a quantity that describes the error in the signal coming out of the sensor, and the noise of the so-called dark current, which is generated in light sensors when there is no photon, is added. The green part shows the addition of the noise caused by the random movement of electrons in the conductor, resulting in voltage changes. The last part is the addition of the noise caused by quantization in an analogue-to-digital converter.

2.3 Noise2Void

Deep neural networks are traditionally trained with two sets of images. One set of images contains pure images, and these images are taken as the absolute truth and used

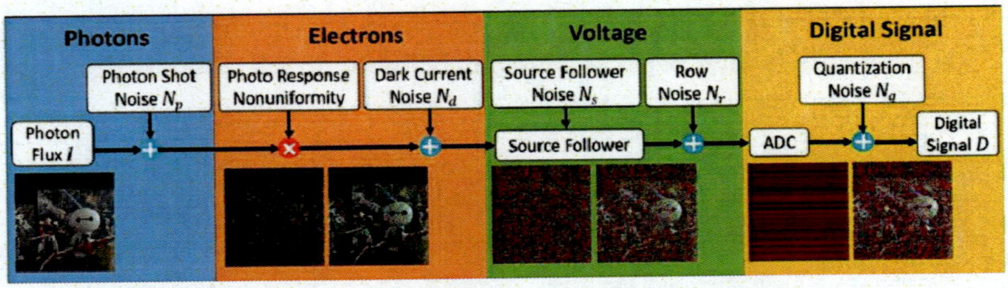

Fig. 3. Visualization of noise and display of the final image in individual stages [2].

as reference in training. The second set of images contains images with noise, which we then want to remove. Sometimes it is difficult, expensive, or even impossible to obtain clean images with which to train a network, and such networks have their limitations. Due to these limitations, Lehtinen et al. [3] have proposed a model of a network that trains with two noisy image sets. These two sets do not contain the same images, but the noise in the images of the first set is different from the noise in the images of the second set. Such an approach is called "Noise2Noise". This is great if a similar can be used image multiple times and when retrieving these images, different noises due to random physical processes are achieved, which should be the only difference between the images of the two sets. Krull et al. [4] have even fewer requirements, so they only need one set for training, a set of images with noise. They realise that it is sometimes impossible to obtain two images that differ only in the noise, or even that it is generally impossible to obtain two images. They see an application in medicine or laboratory imaging where it is important to remove noise from the one available image of which there is no pure version or a version with different noise. They point out that their version is sometimes less successful than previous solutions, but that the difference is very small. This difference is understandable because their network simply works with less available information. The model has been called Noise2Void [4].

2.4 Multi-level Wavelet-CNN for Image Restoration

When developing neural network models, scientists often have to decide between how long to run and train the network and how much detail they want to get in the processed image. To increase the detail of an image, researchers often make their networks deeper, have multiple layers, or increase the size of the filters applied to the image during training. Both of these methods result in an increase in program duration. In their work, Liu et al. [5] try to eliminate this problem, at least partially, by presenting their new model that preserves textures in processed images very well and has a high PSNR value. The execution time in the proposed architecture is clearly not the fastest so far, but the goal is to achieve a balance between preserved features and execution speed. The main idea is to use the discrete wavelet transform, which is similar to the Fourier transform, but instead of representing the signal as a sum of infinite sine waves, they use signals of short duration at specific times of specific frequencies. Such a transform is reversible. Its model can be used not only to remove noise, but also to increase image resolution and eliminate errors in JPEG compression.

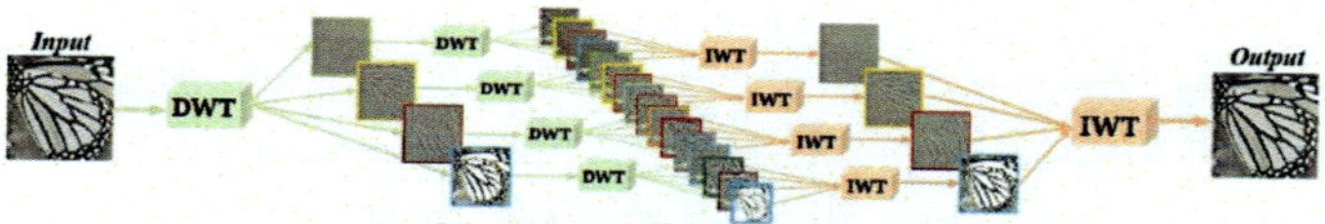

(a) Multi-level WPT architecture

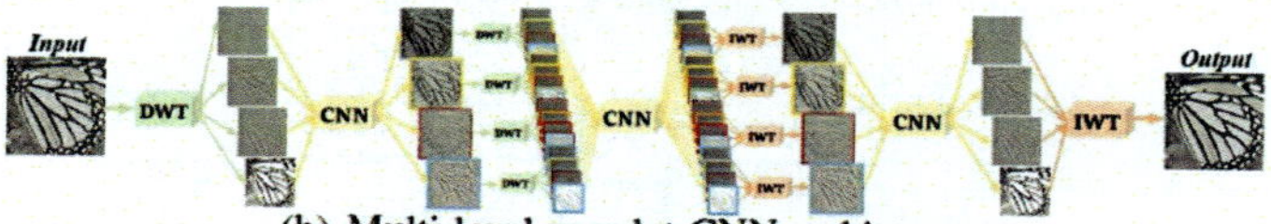

(b) Multi-level wavelet-CNN architecture

Fig. 4. Display of discrete wavelet transformation (a) and proposed architecture (b) [5].

Four images are obtained representing the previous image by applying the discrete wave transformation and forming a plane. On these new images, the transformation can be applied again to obtain four more smaller images (Fig. 4 (a)) [5]. The resulting images are important features of the original image on which the network is trained. Figure 4 (a) also shows the reversibility of the discrete wave transform. The proposed architecture is based on the U-net architecture with the use of the discrete wave transform instead of the originally proposed methods for image transformation. After each transform application, a convolutional neural network is generated (Fig. 4 (b)).

2.5 Deep Image Prior

Ulyanov et al. [6] believe that a deep neural network can recognize image features without training and that the success of the network depends on adapting the network architecture to the data. In the trained networks, the network parameters are set to random values and changed by computing information from a large number of images. The trained networks perform well on images similar to those on which they were trained. The proposed model uses a single image to read new parameters from randomly set parameters and obtain a new, improved image in terms of noise reduction or resolution increase. The authors use the structure of the network as a source of information and show very good results without requiring a large number of images to train.

Each iteration of the neural network searches for the smallest difference between the image obtained by the network and the initial image of the network with random parameters. By reducing these differences, the obtained parameters are used in painting a new image until a satisfactory result is obtained.

x_0 denotes the initial image, f_θ(z) denotes the obtained images with different mesh parameters, Z denotes the fixed tensor on which new images are painted on. E denotes the differences between the images, and θ denotes the network parameters [6, 7]. This is illustrated in Fig. 5.

3 Theoretical Background and Technologies Used

3.1 Deep Learning

Convolutional neural networks Convolutional neural networks show very good results in detecting objects on images, removing noise or similar problems in image processing.

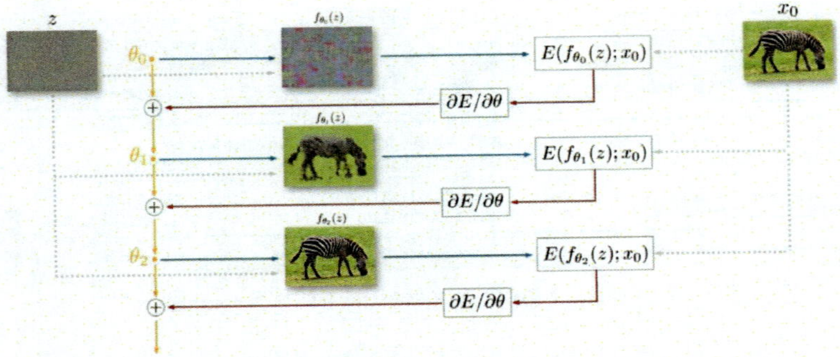

Fig. 5. Basic idea of the model [6].

They are based on MLP type networks with some modifications. They allow very deep networks in a way that eliminates some of the problems of deep MLP networks, such as overfitting to the data and a large number of adjustable parameters, speeding up their training. They take their name from the linear mathematical operation of convolution, which replaces matrix operations during training. Such an operation is denoted by $*$. For continuous functions (signals), it is calculated using Equation 1. The function f is often referred to as the input, the function g as the kernel of the operation, and the output of the operation as the feature map. In convolution, one of the functions is rotated and shifted by the parameter τ. The operation is commutative and is shown in Fig. 6, where the input is the input and the kernel is the kernel.

$$(f * g)(t) = \int_{-\infty}^{\infty} f(\tau)g(t-\tau)d\tau \tag{1}$$

Since computers do not work with continuous but with discrete values, the discrete convolution is more interesting in this case and is represented by Eq. 2.

$$(f * g)(t) = \sum_{-\infty}^{\infty} f(\tau)g(t-\tau)d\tau \tag{2}$$

When processing an image as an input, there is no one-dimensional function, but a convolution on a two-dimensional image with a two-dimensional core. Image is deoted as J, and the nucleus as K; the convolution is shown by Eq. 3 [8].

$$(I * K)(i, j) = \sum_{m}\sum_{n} I(m, n)K(i-m, j-n) \tag{3}$$

Intuitively, neurons can be thought of as pixels of an image arranged in rows and columns. Each neuron receives the intensity of a particular pixel as input. As in MLP networks, these input neurons are connected to the next hidden layer, but not together; instead, a particular neuron from the next layer is connected to a smaller number of neurons from the input layer (Fig. 7). The local receptive held represents the fixed

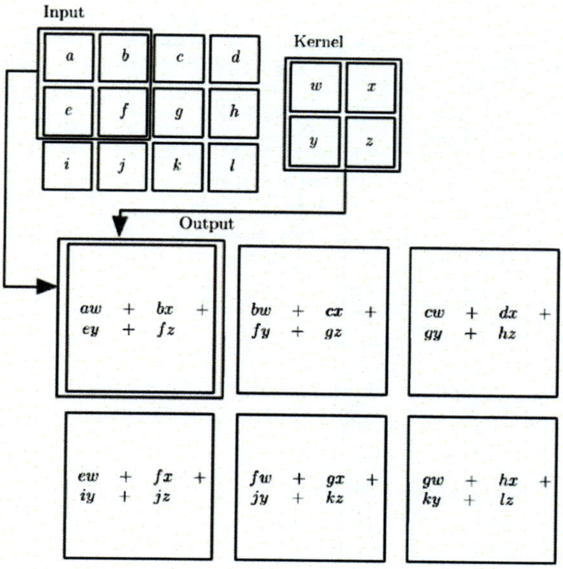

Fig. 6. Convolution [8].

number of neurons to which the neuron from the first hidden layer is connected, and is moved around the image for a given step so that each neuron from the first hidden layer studies a part of the image bounding this field.

All neurons in the first hidden layer have the same threshold and equal weights for connections with the previous layer. This allows each hidden layer to recognize a particular feature that may be located anywhere in the image. The mapping between the input neurons and the next hidden layer represents a feature map, and the common thresholds and weights are called the kernel or filter. A single convolutional layer is defined with multiple feature maps [7].

Each layer of the convolutional network consists of different operations at the entrance of the respective layer. After the described convolution, the information passes through the activation function of each neuron, and then the output of this activation function is processed by aggregating values. This aggregation allows only the essential information for daytime processing to be stored in the rest of the network, further reducing the parameters required to define an individual network. For convolutional meshes, it is often important to identify whether an image contains the desired features, rather than exactly where they are located in a pixel. The aggregation layer processes the output of the individual feature maps and describes whether the desired feature is located in a particular part of the image. An example of an aggregation would be max-pooling, an association where a layer detects the maximum outputs in a particular part of the image. They allow translation independence, in that translating certain features does not significantly change the outputs of the aggregation layers, and they also allow independence from different feature transformations by processing outputs from different

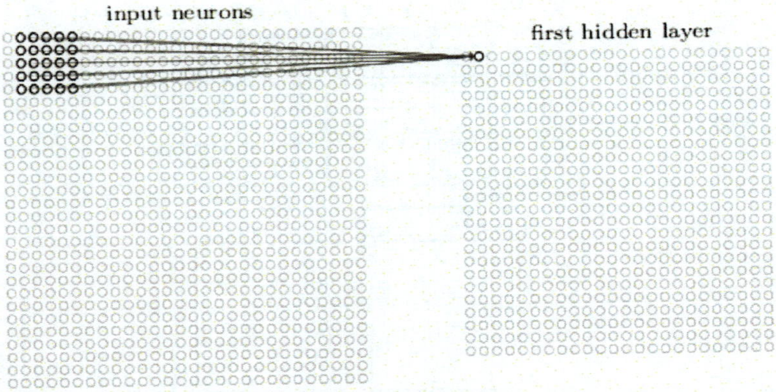

Fig. 7. Intuitive representation of convolution [7].

convolutional layers that are parameterized differently, in which case they would need to associate fewer layers than convolutional layers (Fig. 8) [7, 8].

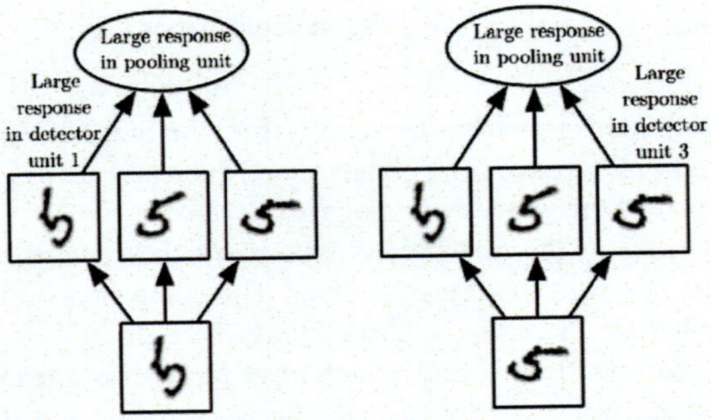

Fig. 8. The aggregation layer allows independence from different feature transformations [8].

The convolution layer is not rotation-independent, so different parameters are required for each number rotation. The last row of Fig. 8 shows two examples of the number five rotated differently, and the row above shows the convolution layer that detects a particular rotation of the number. The left rotation of the number five was detected by the first convolutional layer, while the right rotation was detected by the third convolutional layer. The positive thing about the association layers is that a separate layer is not required for each rotation, but a good response to each of the rotations indicates that the number five is in the given mapping. This simplifies the network; the next number of neurons in the network decreases.

Autoencoder Autocoders are convolutional networks that aim to map input and output almost identically and train like any other feedforward network. They consist of a part that encodes data and a part that decodes data. An autocoder that learns to literally map input to output has learned to simulate an identity function that is not useful. Their purpose is to learn the useful features of images, which is accomplished by constraining their architecture so that they are unable to learn the simple function specified. The constraints force the network to ignore some of the input's information, which is useful in denoising. By gradient descent, autoencoders minimize the error function like general neural networks, but with different parameters, resulting in a denoising autoencoder (DAE) that minimizes the difference between output (processed image with noise) and presented image without noise. DAE thus has the task of removing noise instead of blindly copying the input [4, 8].

U-Net architecture Ronnenberger et al. [9] present a fast network that makes good use of the input images and provides good query of the required features and their localization.

The mesh consists of two parts: a narrowing part and a widening part, which is an innovation compared to other solutions. The narrowing part follows the usual convolutional network with convolution and feature aggregation based on max-pooling aggregation with a step of two. After each feature aggregation, the number of feature maps is doubled so that the context can be passed to other layers. The activation function is the ReLU (rectified linear unit) mentioned above.

The expanding part also follows the usual convolutional network, but instead of merging, it uses inverse convolution, which instead of compressing features, expands the compressed information from the previous layers and halves the number of feature maps. Each inverse convolution adds information from the corresponding constriction layer to the current layer, introducing information about the position of features and creating an architecture reminiscent of the letter U (see Fig. 9). The blue arrows show the convolution with 3 × 3 kernel, the gray arrows add previous matching layers, the red arrows compress the features, the green arrows convert the inverse convolution, and the light blue arrows convert the convolution with 1 × 1 kernel.

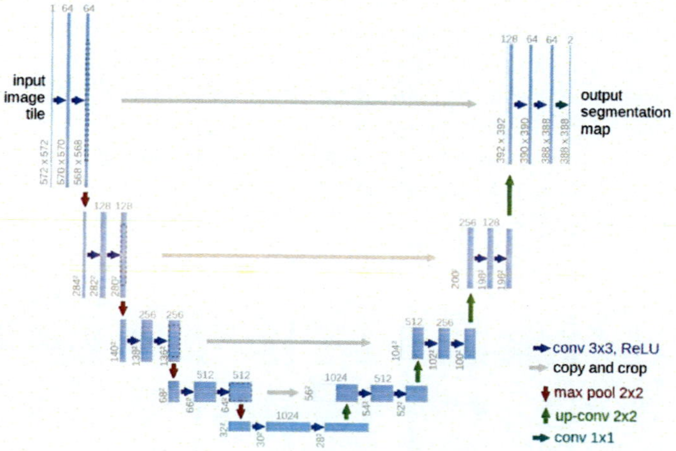

Fig. 9. U-Net architecture [10].

3.2 Keras

Keras is a Python application interface that allows users to easily implement deep learning algorithms. The authors of Keras believe that the principles of Deep Learning are simple and that their implementation should be simple as well. Learning how to use this application user interface is very easy, especially for some simpler projects, the interface supports even the most complex ideas, which of course affects the usability. Basic predefined modules can be used or new modules can be created as needed, Keras is very customizable. It enables engineers and scientists to turn their ideas into tangible results faster. It supports saving already trained networks for use on other computers or platforms, as well as training on processors, graphics cards, and other hardware [11].

4 Image Noise Removal

4.1 Chosen Model

The architecture of the model is based on the U-Net architecture explained in Sect. 3.1, with Fig. 9 showing the architecture of the original work. There are 5 levels of the descending part of the net with 4 associations of features starting from 64 feature maps. When doubled at each level, the left part of the letter U ends up with 1024 feature maps, and such a network has about 31 million variable parameters in total, which challenges the limitation of hardware and execution time. If the U-Net architecture is adjusted so that the initial feature map contains 32 instead of 64, the number of network parameters changes drastically to about 7 million, and such a network can be trained much faster with fewer computing resources. Such an adapted architecture was used to implement autoencoders and remove noise with blind padding.

If the number of feature maps is reduced further, e.g., to 16, it is clear that the number of customizable mesh parameters would be reduced again to about 2 million parameters. However, such a network slows down the error function values much more slowly; it requires more epochs to train and achieves worse results for the same values of the additive white Gaussian noise and salt and pepper noise parameters. Several different initial feature maps were tested. Empirical data suggest that as the size of the network is reduced, the error function values start at a higher number, and as the network remembers fewer features, the quality of the reconstructed images is lower.

Fig. 10. Noise removal with different starting number of feature maps: 16, 32, 64.

Noise removal is shown in Fig. 10 with a varying number of feature maps, with noise-free images in the first row, images with noise in the next row, and noise-removed images in the last row. In the first part of Fig. 10, it was found that the noise removal is much worse than in the other two parts of the image where the noise is removed uniformly. All 3 networks were trained with the same amount of noise and with the same number of levels. Figure 10 confirms the choice of 32 feature maps as the network parameters, since the noise removal is sufficiently good and less computational resources are required.

The CIFAR-10 and MNIST image databases together contain 110,000 training images. Computing the error after each image propagated through the network would take a long time, so the error is computed after a single series of multiple images. Increasing the number of images in a series speeds up the runtime, but the computer hardware requirements are also higher; the training requires more memory. The number of 150 images in a series during the network training was chosen, and such an approximation to the actual errors does not have a great impact on the training result.

The network can be trained with only one of the two image databases listed, and the results obtained are of course interesting. If fewer images pass through the net during training, you train faster. With the same amount of noise and the same number of training epochs, two models were created, one for each of the noise types.

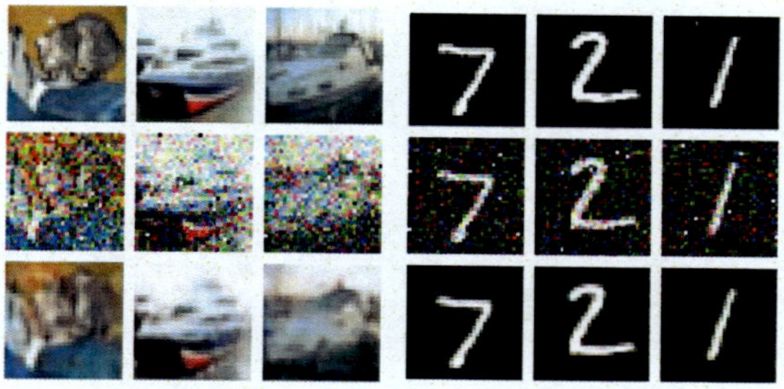

Fig. 11. Noise removal by a model trained on the CIFAR-10 image base.

The model trained on the CIFAR-10 image database removes the standard noise from images from both databases, but the images with noise removed from the MNIST database are no longer completely black and white (see Fig. 11). Noise removal based on this model is satisfactory. When training with images from the MNIST database only, the situation is completely different (see Fig. 12). The denoised images from the MNIST database are completely black and white, like the original image, but the images from the CIFAR-10 database are also colorless, unlike the original images. A network trained in this way does not give satisfactory results because the network has not been exposed to color images. The model for this paper is trained on both image databases because it produces the best results for images from both databases.

It remains to choose the initial amount of the two types of noise.

The parameters that determine the strength of the noise are the standard deviation (Gaussian distributions) σ for additive white Gaussian noise and the probability (pixels that have a minimum or maximum value) p for salt and pepper noise.

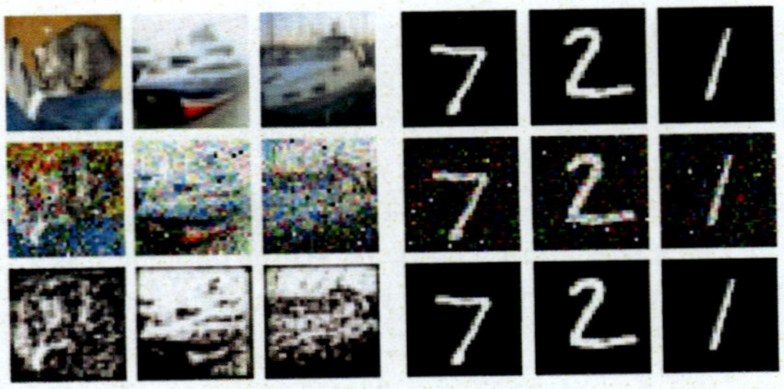

Fig. 12. Noise removal by a model trained on the MNIST image base.

Figure 13 shows the different noise intensities in the images from CIFAR-10 and MNIST. Images from the CIFAR-10 database are much less resistant to noise than images from another database due to their greater complexity, having fewer edges to detect, and being black and white. It was found that the image of the plane is difficult to detect already at the third increase of the noise parameters, while the image of the number 0 is detectable at all noise intensities. Parameters $\sigma - 30, p - 0.04$ were chosen for network training.

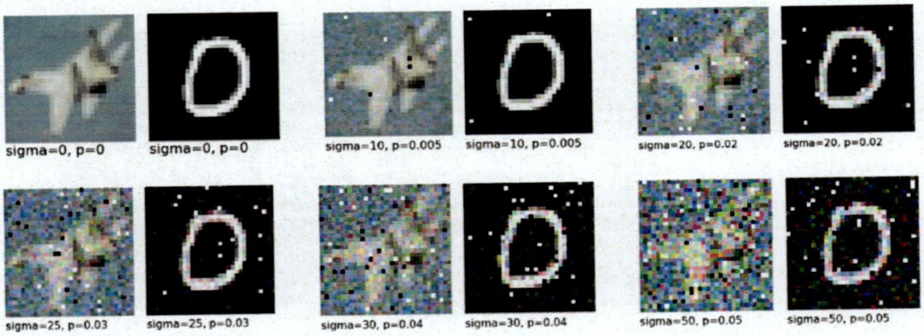

Fig. 13. Various intensities of σ and p.

4.2 Experimental Results

The network was trained for 20 epochs with a lot size of 150. Noise reduction is first shown for images with $\sigma - 30$ and $p - 0.04$ as selected for training the network.

The noise removal from the CIFAR-10 image database (first three thumbnails) is shown on the left in Fig. 14, and the noise removal from the MNIST database is shown on the right. The thumbnails show, in this order: the original image, the image with noise, and the image with noise removed. It can be seen that the original image and the image with noise removed are not completely identical. This discrepancy is reflected in a PSNR of about 16 dB for noisy images compared to the original images. The PSNR for the denoised image from the CIFAR-10 database is 29.12 dB and from the MNIST database is 32.22 dB.

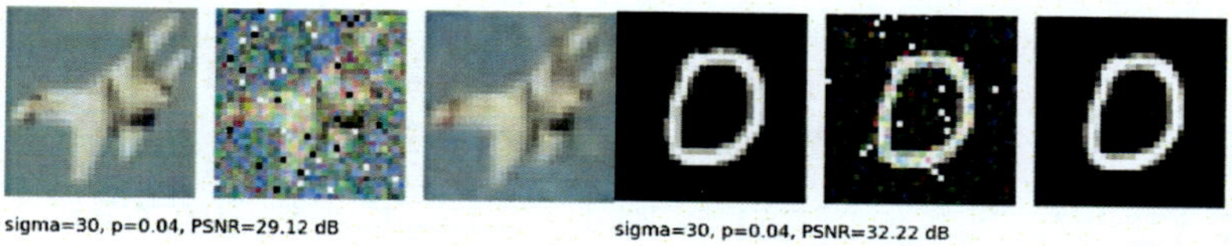

Fig. 14. Noise removal with $\sigma - 30$ and $p - 0.04$.

The improvement after denoising is clearly visible and is shown numerically. The images from the MNIST database are more resistant to noise due to their simplicity and are easier to denoise, which is evident from the reported PSNR values.

Removing only one of the two types of noise listed was also successful. Figure 15 shows the removal of only additive white Gaussian noise, and Fig. 16 shows the removal of only salt and pepper noise.

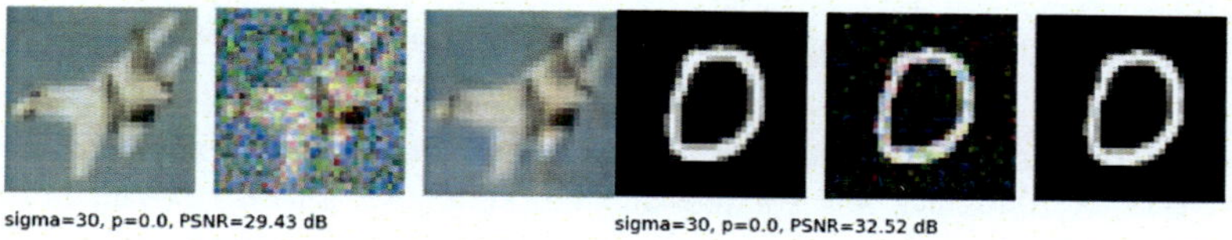

Fig. 15. Noise removal with $\sigma - 30$ and $p - 0.0$.

The PSNR between the original and the image with Gaussian noise is about 18 dB for images from the CIFAR-10 database and about 22 dB for images from the MNIST database. The PSNR for the image with Gaussian noise removed is better than the PSNR for the image with both noises removed.

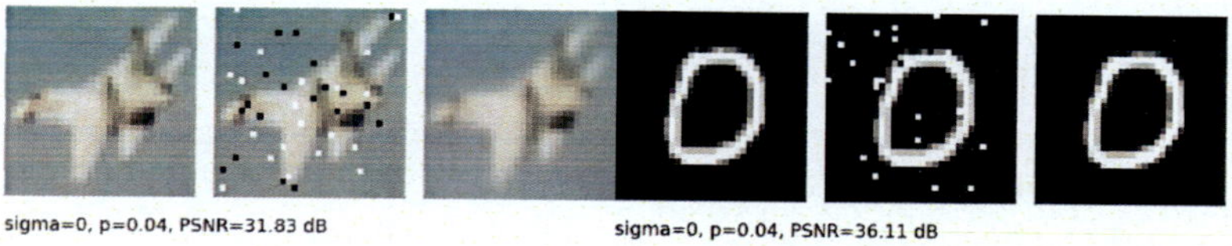

Fig. 16. Noise removal with $\sigma - 0$ and $p - 0.04$.

When only salt and pepper noise is used, the PSNR increases from about 20 dB and 17 dB to 31.83 dB and 36.11 dB for CIFAR-10 and MNIST basis images, respectively, which is an even larger increase compared to the previous examples. The network removes both noises well, both together and individually, regardless of the fact that the images were not exposed with the application of individual noise during training.

The question arises as to what results the network achieves with different noise parameters than the examples in Fig. 13. The noise from the thumbnails left and right (smaller and larger) was selected from the noise selected for training (thumbnails from Fig. 13). These are $\sigma - 25$, $p - 0.03$ and $\sigma - 50$, $p - 0.05$.

If the noise is reduced compared to the noise during training, the PSNR increases (see Fig. 17), otherwise the PSNR decreases (see Fig. 18). The difference can be seen in the appearance of the images themselves. Deep networks do not literally learn where the noise is in the image (that would be difficult since noise is always random), but they learn what part of the image we consider the original part of the image, what noise was added, and how to remove it. The network has seen only certain noise intensities during training, while it has not seen others, and it gives good results even when the parameters are changed. When much more noise is added than it was trained to see, the network only partially removes the noise, but is still relatively successful.

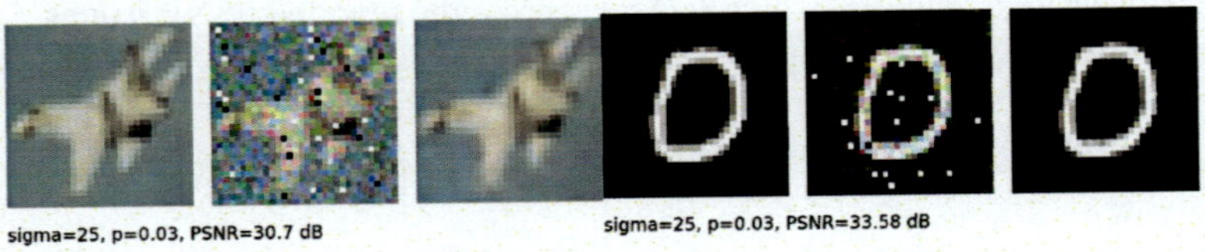

Fig. 17. Noise removal with $\sigma - 25$ and $p - 0.04$.

It is noticeable that the network does a worse job of removing noise based on the CIFAR-10 images, while it does an almost complete job of removing noise based on the MNIST images due to the simplicity of these images as mentioned earlier. In any case, the PSNR improves over the image with noise.

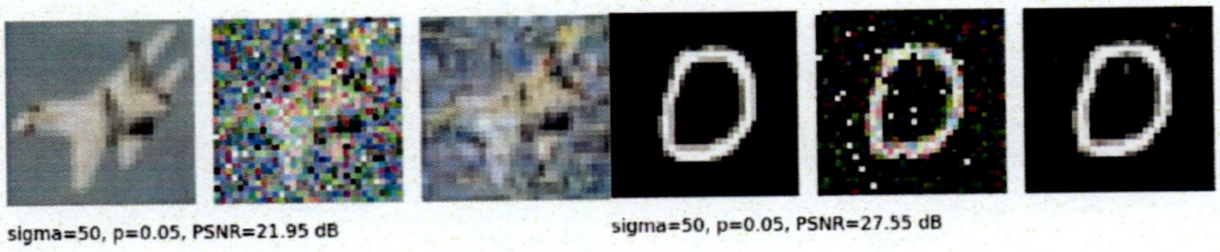

Fig. 18. Noise removal with $\sigma - 50$ and $p - 0.05$.

Table 1 shows the results obtained for both image databases together. The green color indicates the experiment with the images whose σ and p are equal to the one used in the training. The results of the other experiments can be compared with this first experiment. An experiment with reduced noise is highlighted in blue, and an experiment with increased noise is highlighted in red. The previous conclusions are clearly visible at one point. Experiments without a certain type of noise are shown colorless. The first PSNR value indicates the difference between the original image and the image with noise reduction, the second the difference between the original image and the image with noise. Comparing these two PSNRs, a quantitative improvement in image quality can be seen in each of the experiments.

4.3 Application of the Model

The beauty of Deep Learning is the application of the model to similar problems, but with which the network has not yet encountered. Noise removal from a 3088 × 3088 image taken with a mobile device was demonstrated. Since the network expects a tensor (32,32,3) as input, rather than (3088,3088,3), the image is adjusted in two ways. The first method reduces the image to a 32 × 32 image, which loses a lot of image detail, but still produces a new image that the network has not yet seen. The second method is to split a higher resolution image into 32 × 32 parts and finally combine them into a whole image. The noise used is the noise selected during training.

Table 1. Results obtained for certain σ and p.

Image Database	σ	p	$PSNR_{\upsilon}$	$PSNR_{\upsilon}$
CIFAR-10	30	0.04	29.12	16.06
	30	0.00	29.43	18.81
	0	0.04	31.83	19.67
	25	0.03	30.70	17.54
	50	0.05	21.95	13.25
MNIST	30	0.04	32.22	15.93
	30	0.00	32.52	21.62
	0	0.04	36.11	17.31
	25	0.03	33.58	16.42
	50	0.05	27.55	13.49

Fig. 19. Noise removal from the new image with reduced resolution.

The resolution of the acquired image was reduced, noise was added, and this noise was eliminated (see Fig. 19). From about 16 dB for PSNR, the improvement is 23.26 dB. The network did a good job of removing the noise on the previously unseen image of a rose.

The image was split into smaller parts, noise was applied to each part, and then removed. The network does a good job on this example as well. Details can be seen in Fig. 20, where the drawbacks of this approach can be seen, but the results are certainly better than expected. Small images can be seen that make up the higher resolution image in the removed noise image, but these results are considered good since the network is not trained to remove noise from such images.

A PSNR of about 16 dB results in 31.47 dB for the entire image shown in Fig. 21. Without details, it is difficult to see the difference.

Fig. 20. High resolution image details.

Fig. 21. Original image (left) and denoised image (right).

5 Conclusion

Additive white Gaussian noise and salt and pepper noise were successfully removed from images in the CIFAR-10 and MNIST image databases. It has been shown that regardless of the amount of added noise, the network eliminates it at least partially. Image quality is multiplied in all examples. Deep neural networks, despite their complexity, apply well to this problem. Understanding and applying deep neural networks in detail is not an easy task, but it has been shown here to be worthwhile. The deeper we get into the "why" and "how" questions, the more decades of human research and technological development have brought us to the present day. Using the Keras API to implement a neural network model is far simpler and allows anyone interested to model and train at least a simple network and perhaps encourage that person to delve deeper into the world of neural networks. Even complex mathematics does not give us answers to determine certain parameters of the network, but one can arrive at an optimal solution with various trials, and with experience these trials become more and more meaningful. We have determined some parameters empirically and thus gained experience that we will use in further work with neural networks.

Conceptual Design of Smart Furniture Objects Cluster

Domagoj Bilandžija(✉), Damir Blažević, Tomislav Keser, and Andrej Živković

Faculty of Electrical Engineering, Computer Science and Information Technology Osijek, Osijek 31000, Republic of Croatia
domagoj.bilandzija@ferit.hr

Abstract. Given the prevalence of smart furniture and its representation in everyday life, it is logical to expect that smart furniture will become a common part of equipment in accommodation and tourist facilities. Having in mind the needs of the users (e.g. guests), a system has been developed that provides a pleasant experience of using it and content that enriches the stay in the facility. Connecting to a mobile application using Bluetooth technology, the user accesses and manages the system in terms of playing multimedia or selecting the level and/or colour of lighting, with the possibility of fast wired or wireless charging of the mobile phone. The idea of this paper is to use the presence of the system in each room and through it, in addition to the already mentioned features, expand the range of services that can be provided to users. Networking systems with the facility's network infrastructure creates a system of interconnected objects (IoT) that can be jointly managed for the purpose of maintaining, upgrading, sending entertainment and general information or security information to users. This paper presents the concept of IoT smart furniture system that can be used for these purposes.

Keywords: Smart furniture · IoT · ESP32 mesh topology

1 Introduction

Smart building is usually described as an embedded system in the building which is designed to facilitate or completely replace occupants in carrying out daily indoor routines [1]. According to Home Automation Association (HAA), smart building is defined as a processes that uses various methods or appliances to enhance everyday human life and make it more safe, efficient and comfortable [2]. Smart building system is closely associated with the Internet of Things (IoT). IoT is a system of interrelated physical devices which collect and share data over a wireless connection without human intervention [2].

In order to become more competitive, furniture industry are continuously taking care of users needs and desires by adapting to them. In such manner, and with evolution of IoT and artificial intelligence (AI), innovative furniture products are developed. Thus, the classical concept of furniture has changed from just basic objects that humans use

D. Blažević et al. (Eds.): OTO 2022, LNNS 592, pp. 18–24, 2023.
https://doi.org/10.1007/978-3-031-21429-5_2

Of course, these types of noise can be eliminated by changing the architecture with even better results, and the changes require the knowledge gained by working in this area. Some of the different architectures that show very good results are presented in this paper.

References

1. Anwar, S., Barnes, N.: Real image denoising with feature attention. In: Proceedings of the IEEE/CVF International Conference on Computer Vision, pp. 3155–3164 (2019)
2. Wei, K., Fu, Y., Yang, J., Huang, H.: A physics-based noise formation model for extreme low-light raw denoising. In: Proceedings of the IEEE/CVF Conference on Computer Vision and Pattern Recognition, pp. 2758–2767 (2020)
3. Lehtinen, J., Munkberg, J., Hasselgren, J., Laine, S., Karras, T., Aittala, M., Aila, T: Noise2Noise: Learning Image Restoration Without Clean Data (2018). In: arXiv preprint arXiv:1803.04189
4. Krull, A., Buchholz, T.O., Jug, F.: Noise2void-learning denoising from single noisy images, In: Proceedings of the IEEE/CVF Conference on Computer Vision and Pattern Recognition, pp. 2129–2137 (2019)
5. Liu, P., Zhang, H., Zhang, K., Lin, L., Zuo, W.: Noise2Noise: Multi-level wavelet-CNN for image restoration. In: Proceedings of the IEEE Conference on Computer Vision and Pattern Recognition Workshops, pp. 773–782 (2018)
6. Ulyanov, D., Vedaldi, A., Lempitsky, V.: Deep image prior. Int. J. Comput. Vision **128**(7), 1867–1888 (2020). https://doi.org/10.1007/s11263-020-01303-4
7. Nielsen, M.A.: Neural Networks and Deep Learning, vol. 25. Determination Press, San Francisco (2015)
8. Kriesel, D.: A Brief Introduction to Neural Networks. http://www.dkriesel.com/en/science/neural_networks. Last Accessed 5 Dec 2021
9. Ronneberger, O., Fischer, P., Brox, T.: U-net: Convolutional networks for biomedical image segmentation. In: International Conference on Medical Image Computing and Computer-Assisted Intervention, pp. 234–241 (2015)
10. Goodfellow, I., Bengio, Y., Courville, A.: Deep Learning. MIT Press (2016)
11. Keras: https://keras.io/about/. Last accessed 12 Feb 2022

to furniture units that are able to sense, identify and interact with users to meet their needs and desires [3]. According to that, smart furniture is a relevant development and tendency in the furniture industry [4].

Among others, smart furniture is regularly one of the subsystems that are integrated in a smart building system. The possibility of communication is one of the crucial device functionalities for smart furniture that is needed to be controlled [5]. Here, communication is two folded. Firstly, communication in terms of data transfer between end-user and smart furniture object and secondly, communication in terms of data transfer between two or more smart furniture objects. When it comes to communication, it is always important to keep in mind possible device compatibility issues as well as security issues, to ensure privacy and safe usage of equipment [5].

In our observed case, nightstands that are part of smart furniture, are distributed in rooms of some tourist facility. Such nightstands are equipped with ESP32-WROOM-32D microcontroller which could be connected to an occupant's mobile phone application via Bluetooth technology. In order to connect each smart furniture nightstand which are spatially distributed depending on the construction of the building, ESP MESH topology is used.

This paper is organized as follows: the central part of the paper which deals with mesh topology consist of several subsections. Namely, the comparison between traditional Wi-Fi mesh and ESP mesh, detailed description of the ESP-Wi-Fi-MESH topology, data transmission and application of ESP-Wi-Fi-MESH topology, respectively. At the end, the concluding remarks are derived.

2 Mesh Topology

Mesh networking is a local area network topology. It consists of nodes which are directly, dynamically and non-hierarchically connected to as many nodes as possible. Nodes cooperate with one another to efficiently route the data from and to clients. Lack of dependency on one node allows that every node can participate in transferring information.

A wireless mesh network is a communication network built of radio nodes communicating with each other in mesh topology. Every node can communicate with other nodes in range and, essentially, they form an intelligent grid of access points all of which can communicate with each other to smartly route traffic through the network.

ESP-MESH is a networking protocol built on the top of the Wi-Fi protocol and it lets numerous devices (generally known as nodes) that are distributed over a wide area to be mutually connected under a single Wireless Local Area Network (WLAN). ESP-MESH allows multiple nodes to communicate with each other due to being in same WLAN. It is supported by both ESP32 and ESP8266 microcontroller.

2.1 Comparison: Traditional Wi-Fi and ESP-Wi-Fi-MESH

A traditional Wi-Fi network infrastructure, shown in Fig. 1 [6], can be explained as a point-to-multipoint network where only one node is the central node known as the access point (AP). AP is directly connected to all other nodes in network known as stations. This central node is responsible for arbitrating and forwarding transmissions between the stations and also some APs relay transmissions to and from an external Internet Protocol (IP) network via a router. In traditional infrastructure of a Wi-Fi networks every station or node must be in range of the AP in order to connect to it. That causes certain limitation in terms of covering larger areas with signal. Every station receives AP signal whose strength depends on distance between them. Stations closer to AP achieve greater data transmission speed provided with stronger signal. Stations outside of the AP range are unable to connect and achieve any kind of transmission. Also, traditional Wi-Fi networks are easily overloaded as the maximum number of stations allowed in the network is limited by the capacity of the AP.

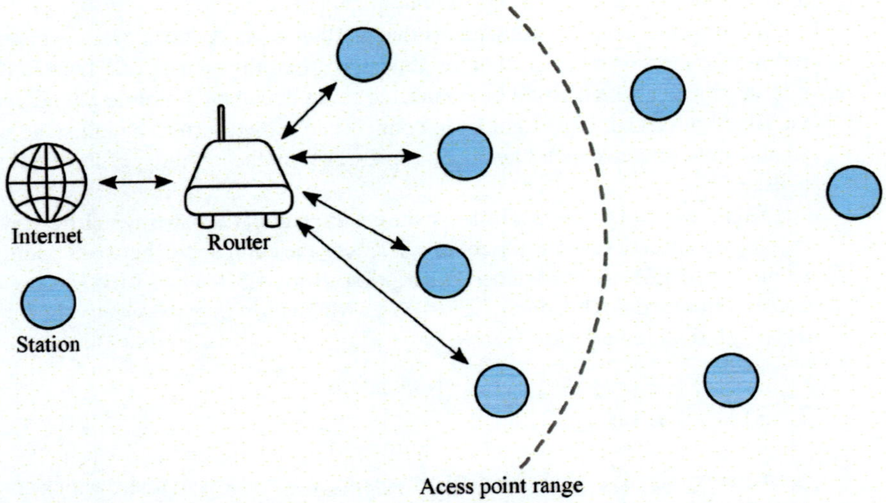

Fig. 1. Traditional Wi-Fi network architecture.

ESP-Wi-Fi-MESH protocol is different from traditional infrastructure of Wi-Fi networks in a way that nodes are not required to connect directly to a central node but, instead of that, nodes are allowed to connect with neighboring nodes and they are responsible to mutually relay each others data transmissions. This allows an ESP-Wi-Fi-MESH network to have wide coverage area as nodes can achieve mutual connection even if they are not in range of the central node. Such general ESP-Wi-Fi-MESH network is depicted in Fig. 2 [6]. ESP-Wi-Fi-MESH network is also less susceptible to overloading as the number of nodes allowed in the network is not limited by a single central node.

2.2 ESP-Wi-Fi-MESH Topology

ESP-Wi-Fi-MESH is built atop the Wi-Fi protocol infrastructure and can be thought of as a networking protocol that combines many individual Wi-Fi networks into a single WLAN. In Wi-Fi, stations are limited to a single connection with an AP (upstream connection) at any time, while an AP can be simultaneously connected to multiple stations (downstream connections). However ESP-WIFI-MESH allows nodes to simultaneously

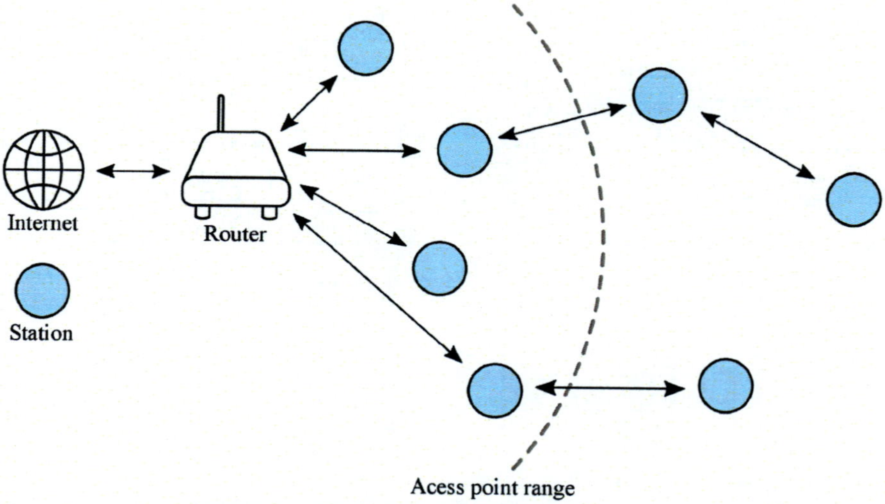

Fig. 2. ESP-WIFI-MESH network architecture.

act as a station and an AP. Therefore a node in ESP-WIFI-MESH can have multiple downstream connections using its softAP interface, while simultaneously having a single upstream connection using its station interface. This feature results in a tree network topology with a parent-child hierarchy and it is consisted of multiple layers.

The root node is the top node in the network and serves as the only connection between the ESP-WIFI-MESH network and an external IP network [6]. The root node is connected to a traditional Wi-Fi router and relays packets to and from the external IP network to nodes of that ESP-Wi-Fi-MESH network. Only one root node within an ESP-WIFI-MESH network is possible and the upstream connection of the root node is exclusively with the router. Root node is presented in both Figs. 3 and 4 [6].

A leaf node is one that is not allowed to have any child nodes (no downstream connections). Hence, a leaf node can only transmit or receive its own packets, but cannot forward the packets of other nodes [6]. Node is always assigned as a leaf node if its on maximum permitted layer of a network. In that way node is unable to form any downstream connections and it is ensured that extra layer will not be added in a network. Some nodes are without a softAP interface (station only) and they are assigned as leaf nodes due to the requirement of a softAP interface to have downstream con-nections. Nodes L, M, N are leaf nodes shown in Fig. 4.

Intermediate parent nodes are nodes that are part of the network and they are neither the root node nor a leaf node. An intermediate parent node must have a single parent node (upstream connection) but can have zero to multiple child nodes (zero to multiple downstream connections). Intermediate parent node has the ability to transmit, receive and forward packets sent from its upstream and downstream connections. In Fig. 4, all nodes marked from letter B to letter J are intermediate parent nodes.

Ultimately, idle nodes are nodes that are not part of network but may optionally join the network. Idle nodes will try to join a network by forming an upstream connection

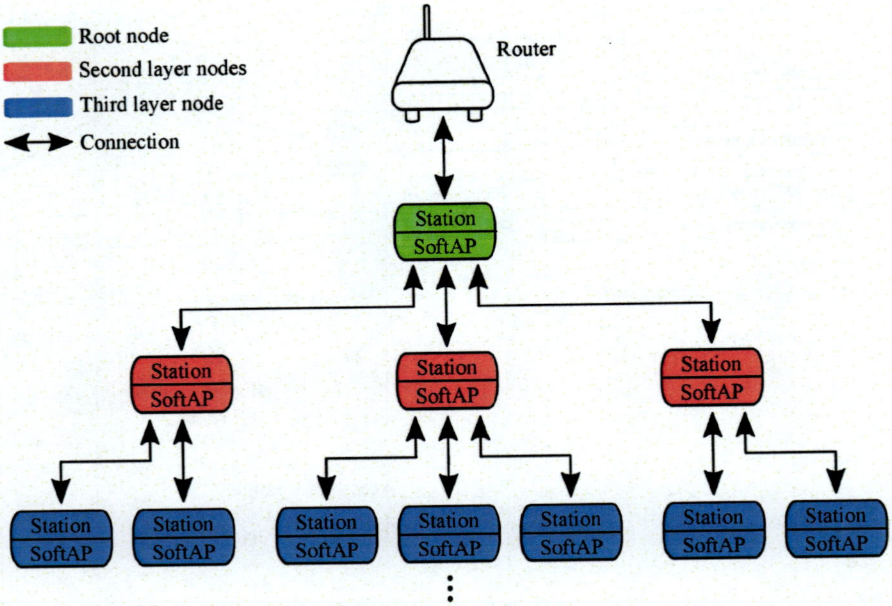

Fig. 3. ESP-Wi-Fi-MESH tree topology.

with an intermediate parent node or by becoming the root node under the favorable circumstances. Nodes marked by K and O, in Fig. 4, are idle nodes.

2.3 Data Transmission

ESP-Wi-Fi-MESH packets are special kind of package that is used for ESP-Wi-Fi-MESH network data transmission. They are contained within the frame body of a Wi-Fi data frame [6]. A multi-hop data transmission in an ESP-Wi-Fi-MESH network will involve a single ESP-Wi-Fi-MESH packet being carried over each wireless hop by a different Wi-Fi data frame [6]. Structure of ESP-Wi-Fi-MESH packet and its link to Wi-Fi dataframe are displayed in Fig. 5 [6].

The Header of an ESP-WIFI-MESH packet consists of the MAC address of the source node (Src Addr) and destination node (Dest Addr). Remaining part of the Header is the Options that consists of information about special types of ESP-Wi-Fi-MESH packets. Special type of packet is a packet originating from the external IP network or group transmission. Second portion of an ESP-Wi-Fi-MESH packet is the Payload and it holds the actual application data which can be raw binary data or data encoded under an application layer protocols such as HTTP, MQTT, and JSON [6].

2.4 Usage

ESP-Wi-Fi-MESH network technology allows us to make multiple products based on ESP32 which are supposed to communicate with each other or receive some sort of

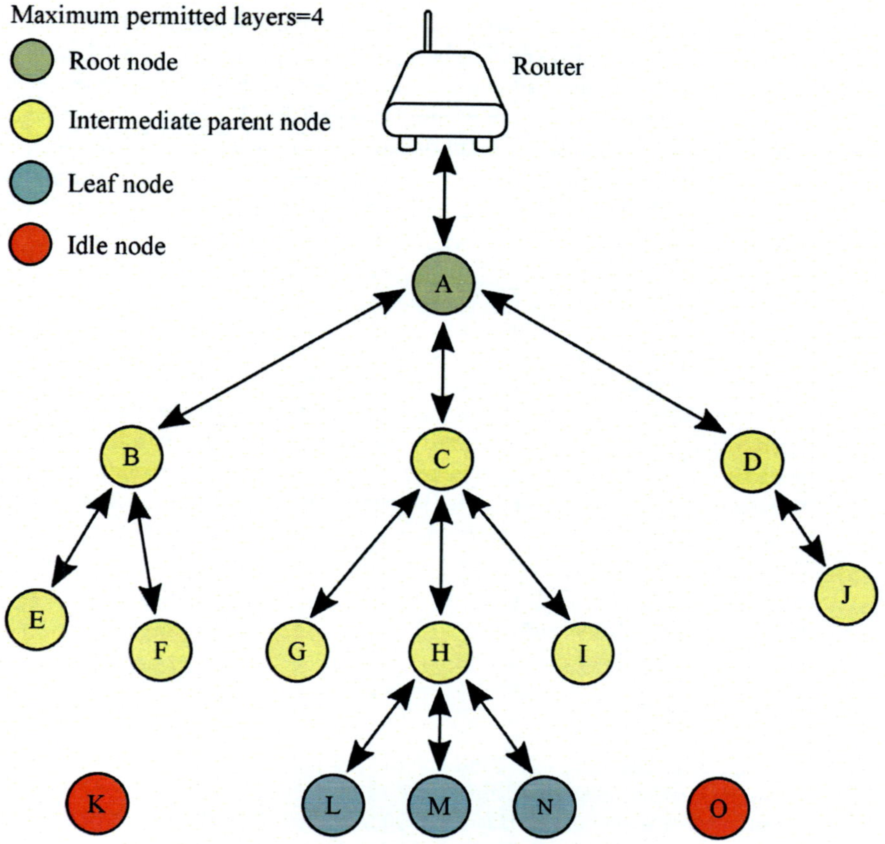

Fig. 4. ESP-WIFI-MESH node types.

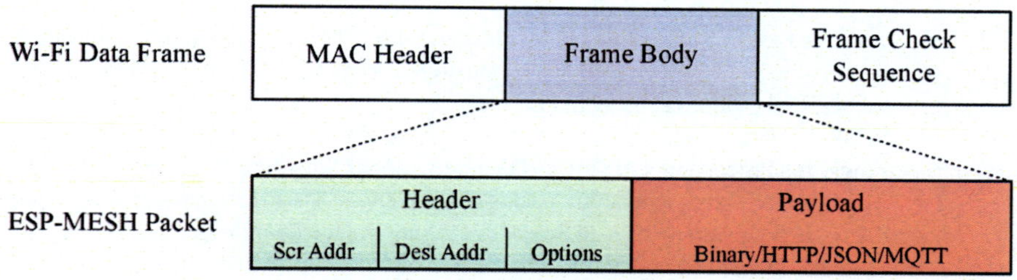

Fig. 5. ESP-WIFI-MESH packet.

global message. In our case of smart furniture there are numerous ways to make use of such technology. Multiple pieces of furniture located in tourist accommodation facilities greatly benefit from possibility of being connected with each other. In bigger hotels it can be challenging to cover all areas with great Wi-Fi signal strength. For example, if there is a router in main hallway there is a possibility that inside of a room or on a balcony, signal strength would be greatly reduced. By having several pieces of smart furniture inside of a room problem could be easily solved without extra investment in additional hardware such as wireless signal range extender or additional routers inside of a room. Piece of a furniture which is receiving strongest signal from main router can transmit information to other pieces of smart furniture inside one room and greatly extend signal range without need of additional devices. Smart furniture equipped with audio features can receive and transmit global message sent via main hotel Wi-Fi.

3 Conclusion

ESP-Wi-Fi-MESH is a very useful technology which has multiple applications that can improve efficiency of existing network hardware and reduce need for extra investment. It is essentially the fundamental technology on which efficient and ubiquitous system of interconnected objects can be realized by using ESP32 microcontrollers. Smart furniture systems are great example as they are part of almost every room of smart buildings which are becoming more and more present in our lives. From small private household ranging to great tourist accommodation buildings, smart furniture systems and their interconnectability are more than able to improve quality of life and enhance experience we expect from piece of furniture either by entertainment features, information feed or important security warnings.

References

1. Syafa'ah, L., Minarno, A.E., Sumadi, F.D.S., Rahayu, D.A.P.: ESP 8266 for control and monitoring in smart home application. In: 2019 International Conference on Computer Science, Information Technology, and Electrical Engineering (ICOMITEE), pp. 123–128 (2019). https://doi.org/10.1109/ICOMITEE.2019.8921287
2. J. Li, Y. Lin, IOT home automation—smart homes and internet of things. In: 2021 3rd International Academic Exchange Conference on Science and Technology Innovation (IAECST), pp. 294–298 (2021). https://doi.org/10.1109/IAECST54258.2021.9695788
3. Mohamed, A.M.S., Ibrahim, N.F., Hassan, M.R.: The impact of smart furniture design on meeting the needs of a limited-space housing. J. Architect. Art. Hum. Sci. **6**, 25 (2021). https://doi.org/10.21608/mjaf.2020.23392.1499
4. Oblak, L., Kropivšek, J., Jošt, M., Zupančić, A.: Furniture development based on internet of things. In: Proceedings of 2017 10th International Scientific Conference WoodEMA: More Wood, Better Management, Increasing Effectiveness: Staring Points and Perspective, pp. 186–191, Czech University of Life Sciences Prague, Prague (2017)
5. Blažević, D., Bukovčan, M., Keser, T., Nenadić, K.: Conceptual design of smart furniture: a case study. In: 30th International Conference on Organization and Technology of Maintenance (OTO 2021). OTO 2021. Lecture Notes in Networks and Systems, vol. 369. Springer, Cham (2022)
6. ESP-WIFI-MESH. https://docs.espressif.com/projects/esp-idf/en/stable/esp32/api-guides/esp-wifi-mesh.html. Last accessed 26 Aug 2022

Organization of a Digital Voting System Based on Blockchain Technology for the Faculty Council

Mirko Köhler, Miljenko Švarcmajer(✉), Ivica Lukić, and Tomislav Stipanić

Faculty of Electrical Engineering, Computer Science and Information Technology Osijek, 31000 Osijek, Croatia
miljenko.svarcmajer@ferit.hr

Abstract. Digital/online voting has great potential to decrease organizational costs and increase voter turnout because voters do not need to be present during meetings to cast a vote, they can vote from wherever there is an Internet connection. Despite obvious benefits, digital voting solutions are viewed with a great deal of caution because a single vulnerability can lead to distrust in the results of voting and the organization itself. Blockchain technology offers decentralized nodes with pear-to-pear verification advantages, security protection characteristics, non-repudiation and digital signatures. In the paper organization of the Faculty Council session, roles of its members, and the current ways of decision-making are described. This paper proposes using Blockchain to replace the existing voting system using e-mail. Security and data integrity of votes is absolutely provided theoretically. The organization of such a voting system is given in this paper. Voting application and resulting session report is proposed.

Keywords: Blockchain · Smart contract · Digital voting

1 Introduction

When it comes to democracy, as well as a democratic society, people immediately think of voting. Although it is already well into the 21st century, paper balloting is still considered the most reliable and certain way to make sure a person has fully exercised their right to vote [1]. In this paper, the concept of organizing a digital voting based on blockchain technology is presented, using a relatively small body such as the Faculty Council voting system. The Faculty Council is a faculty body that makes important decisions and whose work is described in detail by the faculty statute. A more detailed description of the Faculty Council role will be given in Chap. 2. Previously, all Faculty Council sessions were organized physically, at the same location. However, the COVID-19 pandemic has shown that this will not always be possible and a more modern solution was needed without physical contact. As a logical step, in a time when it was impossible to physically meet, e-mail sessions and video conferences were conducted, but there are more modern solutions using blockchain technology which will be presented in this paper.

D. Blažević et al. (Eds.): OTO 2022, LNNS 592, pp. 25–36, 2023.
https://doi.org/10.1007/978-3-031-21429-5_3

The following chapters will describe the organization of the Faculty Council session, roles of its members, and the current ways of decision-making. This is followed by an introduction to blockchain technology, smart contracts and a proposal for the concept of using this technology in decision-making by the faculty council is given. The described concept would enable a free, accessible and transparent way of voting, as well as full confidence in the results with blockchain traceability. However, although modern methods are generally considered to have advantages, they will be accompanied by possible disadvantages that will be discussed.

2 Organization of the Faculty Council Session

The Faculty of Electrical Engineering, Computer Science, and Information Technology Osijek (FERIT) has the Statute. A statute is a collection of rules and regulations used by different companies or institutions. A faculty statute regulates the internal structure, management issues and decision-making [2], and it is the most important document of an institution. According to the Statute of FERIT [3], the two most important bodies of the FERIT are the Dean and the Faculty Council.

2.1 Dean

Among a lot of other responsibilities, Dean has a crucial role in Faculty Council Sessions. The Dean chairs the sessions of the Faculty Council and has the following rights and duties:

- convenes and conducts sessions,
- proposes the session agenda and makes sure that the session takes place as determined agenda,
- maintains order at the session and gives the floor to the speakers,
- refers to the discussion and decision-making of the Faculty Council, prepares proposals, reports, opinions, analyses and other items within the competence of the Faculty Council,
- determines and publishes the voting results,
- takes care of keeping the minutes of the session,
- ensures that the work of the Faculty Council respects the provisions of the Law, the Statute and the general acts of the University and the Faculty,
- signs decisions and general acts adopted by the Faculty Council,
- performs other tasks in accordance with the Law, the Statute of the University and the Faculty, general acts of the University and the Faculty and these Rules of Procedure.

2.2 Faculty Council

According to [3] The Faculty Council consists of all full professor with tenure, full professors, associate professors, assistant professors, two representatives of lecturers elected to teaching positions, two representatives of associates elected to associate titles, one representative of other employees who have a contract of employment with the

Faculty, and student representatives who make up at least 15% of the total number of members of the Faculty Council. The work of the Faculty Council and the manner of decision making, in addition to the Statute, are determined in more detail by the Rules of Procedure of the Council [4]. By default, Dean is a member of the Faculty Council.

The Faculty Council of FERIT is organized according to the rules written in the Statute of FERIT. The main role of Faculty Council members is to attend sessions of the Faculty Council. At them, Faculty Council members participate in discussions and vote on items proposed by the Dean. Responsibilities of the Faculty Council are:

- vote on proposed items,
- makes decisions on academic, scientific, and professional issues,
- elects and dismisses deans and vice-deans,
- adopts the Statute and other general acts at the proposal of the Dean,
- initiates and conducts part of the selection procedure for scientific titles,
- initiates and conducts the procedure of election to scientific-teaching, teaching, associate and professional titles and appropriate positions,
- organizes postgraduate university studies,
- organizes postgraduate specialist study,
- determines the structure of the Faculty,
- decides on the establishment of new organizational units of the Faculty,
- appoints and dismisses heads of institutes,
- proposes to the Dean the Rulebook on the organization of jobs.

The Faculty Council decisions are made by public voting by a majority vote of the members present, except in cases where the Law, the Statute of the University, the Statute of the FERIT or another general act stipulates otherwise. In those cases, the decision must be made by a majority of the total number of members and/or by secret vote.

2.3 Session of Faculty Council

There are four types of Faculty Council Sessions. They can be regular, extraordinary, elective and ceremonial. Elective and ceremonial sessions are not in the interest of this paper. There is no voting at ceremonial sessions, and elective sessions must be held in public. Regular sessions of the Faculty Council are held as a rule once a month and extraordinary sessions of the Faculty Council are held based on the indicated need or justified reason.

As stated before, the Dean proposes the agenda of proposed items to be discussed at the session. The Faculty Council makes decisions by public vote and by a majority vote of the members present, except in cases where the Law, the Statute of the University, the Statute of the FERIT or another general act stipulates otherwise. In those cases, the decision must be made by a majority of the total number of members and/or by secret vote.

An electronic session of the Faculty Council may be convened by the Dean in urgent and justified cases. For valid decision-making at the electronic session, the Dean must submit a decision proposal by e-mail and set a deadline by which voting is not shorter than 24 h, within a working day. The present members of the electronic session are those members who voted in the electronic session and decisions must be made by a majority of all members of the Faculty Council. Voting is done using the official e-mail address assigned to the Faculty Council member by FERIT. After an individual member of the Faculty Council receives a proposed item's, by e-mail, he should vote on them in specific way. The ordinal number of the proposed item should be given first and then after that one of the decisions: "FOR", "AGAINST" or "ABSTAINED" [4]. Electronic session of the Faculty Council does not have an oral discussion. All documents related to the individual decision proposal, as well as the invitation to the session with the agenda, are submitted as attachments in e-mail. That is why it is very important that decisions that could provoke controversy or require discussion are not put on the agenda of electronic sessions.

After the vote, the dean determines whether the individual decision received the required majority of votes and announces the result of the vote. Previously made decisions may be revoked, annulled, or changed by the Faculty Council at one of the next sessions if a new factual situation related to the decision-making is established.

Each new session begins with the adoption of the written Report from the previous session of the Faculty Council. Report of the electronic session of the Faculty Council must be accompanied by a printout of all e-mails sent by members of the Faculty Council containing their votes.

3 Blockchain Based Voting

As explained in the previous chapter, there is a need for electronic sessions and digital voting. Whether it is case of an emergency, or some other justified cases like in last years it was even forbidden to hold public gatherings due to measures introduced in the fight against the COVID-19 pandemic.

The current voting system for electronic sessions takes place via e-mail. Traceability of such voting is less than desirable, messages can end up in spam or some other mail folders, as well as replies to them. As proof of voting, it is necessary to print out all the responses sent by the Faculty Council members and archive them. Messages can be accidentally deleted or forwarded to the wrong address. Voting must be done from an official email given by the faculty in order to be valid.

This paper will present the organization of digital voting using blockchain technology. In order to better explain this proposed voting solution, it is necessary to explain used technologies. To understand why blockchain is used in the voting system at all, it is necessary to explain its basics and advantages like traceability, trust and transparency.

3.1 Distributed Ledger Technologies

Distributed Ledger Technologies (DLT) offers a way to increase trust, traceability, and collaboration within one or more institutions. Most known DLT is a blockchain [5]. Blockchain is a growing record of data in a slightly unique way than with standard databases. Various security mechanisms are also used in blockchain. Every record is stored in the block and when the block is full, it gets its timestamp and a hash of the previous block. The timestamp is proof that the date stored in the blockchain existed when the block was published. While the hash is a mathematical function that turns an arbitrary-length input into a fixed-length encrypted output. As a result, its unique hash will always be the same size, regardless of the original quantity of data or file size involved. Furthermore, because hash functions are "one-way," they cannot be utilized to "reverse-engineer" the input from the hashed result. Even yet, if you perform the same function on the same data, the hash will be the same, allowing you to verify that the data is the same [6]. The main use of blockchain, in the beginning, was cryptocurrencies, but very quickly blockchain attracted the attention of various business sectors. The blockchain is used in supply chain management, healthcare, identity management, and conventual financial services as shown in Fig. 1.

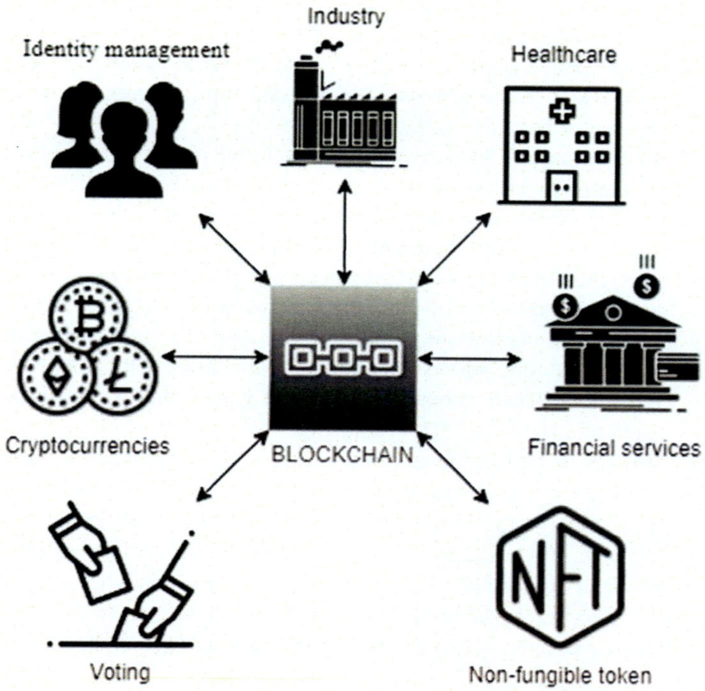

Fig. 1. Blockchain use cases.

3.2 Blockchain Voting

To put it simply, blockchain voting is a variant of electronic voting that adds digital (cryptographic) signatures to make it more difficult to forge votes, due the hashing and distributed consensus to keep votes secure are used. Computer networks use software to agree on the data's arrangement and content. As previously stated, such software is public and accessible to everybody. As the software attempts to check and link received data using hash algorithms that prohibit data modification, users upload fresh data as well as digital cryptographic signatures. As a result, previous data is preserved, unaltered, and available to all users. It is critical to highlight the following qualities in the context of voting:

(1) **The consensus protocol** is a set of rules used by nodes in a network, which leads to the conclusion that is best for the entire network. Although there is no central body for validation and verification, it is considered that every transaction in the chain is fully secured and verified. A consensus or protocol that allows any computer to connect and participate in a network is called a permissionless protocol. The blockchain which uses permissionless protocol is called public and has been used in most of the currently most popular blockchains (Bitcoin, Ethereum, Litecoin, Cardano, etc.). The advent of the permissionless protocol has forced many to reconsider distributed databases where the set of participants is predetermined and limited. These protocols improve error tolerance and may even tolerate some malicious nodes. Such a protocol is called a permission protocol, and the blockchain that uses it becomes private.

(2) **Authentication**. There is no traditional user identity in the blockchain system, the only thing that represents the user is the private key with which he performs digital signing. The user is solely responsible for managing and storing his private key. In case the private key is stolen or lost, the user loses his "identity" on the blockchain.

(3) **Smart contracts**. A smart contract is computer code that is automatically executed at defined conditional events described within the contract. In this way, it is possible to perform much more complex actions on the blockchain than just transferring values from one address to another. With smart contracts, it is possible to create applications such as markets, computer games and various decentralized financial applications.

(4) **Secrecy of transactions**. All transactions on the blockchain are public, at least in most cases. This is one of the key features, that all transactions are transparent and verifiable. In a private blockchain, it is possible to restrict access to read data, which can be useful to limit data leakage but those without access cannot participate and check the blockchain. Some systems use "zero-knowledge proof" to hide details of transactions, participants and amounts. Proof of zero-knowledge reliably shows that a statement is true without revealing why that statement is true [7].

Most research related to blockchain-based electronic voting is related to voting in public elections where anonymity is particularly important [8, 9].

Blockchain-based electronic voting usually looks like this. The electorate, which conducts the election and has a list of all voters, creates a pair of keys, public and private, that will represent that voter. In addition to the voting question and candidates, the period within which they will be able to vote is defined. Voters send their votes signed with their private key before the deadline, after which the election closes and the votes are counted.

Blockchain-based voting scenario schematics, as shown in Fig. 2, is consisting of:

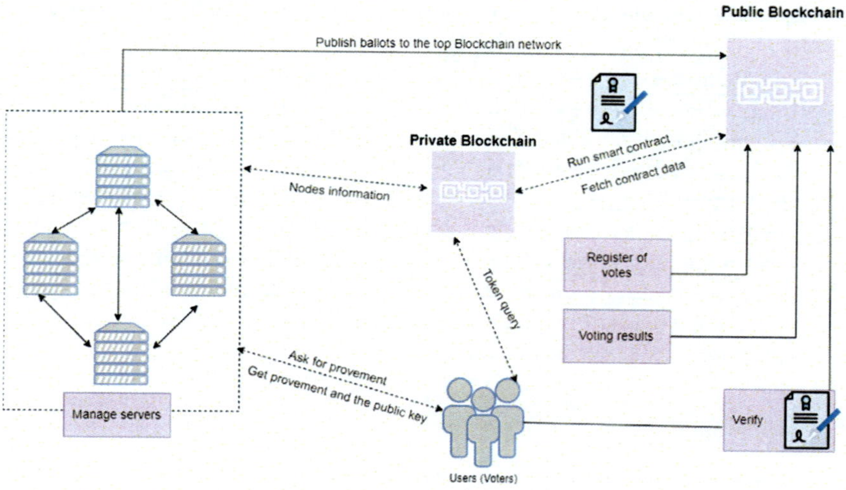

Fig. 2. Blockchain-based voting scenario.

Manage servers: Its purpose is to store node information in the lower blockchain network, broadcast it to the higher blockchain network, and issue certificates to nodes. This enables node authentication as well as the usage of user credentials to access the system.

Blockchain network: The blockchain network in the proposed scenario is made up of many blockchains that work in tandem. This structure enables parallel execution, which boosts the system's overall performance and scalability. Because each node in the private chains has a local blockchain that holds the privacy-sensitive data, the private chains are used to store node information and the voter identity register. After certain voters successfully agree on the transactions, the public blockchain (such as Ethereum) serves to record independent blockchain states across all voters and concurrently process transactions. The transactions that are recorded in the public blockchain are trusted and unchangeable.

Users (Voters): Users are both voters and members of the election committee; they can authenticate and access their wallets using their identity ID. Voters are given a digital token that enables them to cast their ballots. As a result, smart contracts are deployed in the blockchain's public section (Ethereum blockchain).

Smart contract: In the proposed decentralized system, smart contracts are self-executing pieces of code. The contract agreements that allow transactions in the public blockchain network to be tracked are defined by the functions encoded in smart contracts. Each node in the blockchain network can independently run the smart contract to obtain consensus in the proposed scenario, resulting in the establishment of a modular cryptosystem for voting systems [10].

4 The Organization of a Faculty Council Voting System

In this chapter, the organizational structure of the electronic session of the Faculty Council and the voting itself, on the proposed decisions, will be presented. The roles of the system will be presented first and then the components of the smart contracts will be explained. Also, the system architecture will be proposed, and a description of voting applications will be given. One of the most significant differences between the voting presented here and the solutions presented so far in the article lies in the fact that there is no anonymity in this voting system. Voting in political (parliamentary or presidential) elections must be by a secret vote. This is one of the principles of democratic systems. However, in this example, it is required that the vote be public and that it can be determined exactly who voted how.

4.1 Roles

There are three roles in this system. The two, most important roles, we have already explained: Dean and Faculty Council members. The third role in the system is the Secretary of the Faculty (Secretary). The role of the Secretary, within the sessions of Faculty Council, is to perform organizational, professional-administrative, legal, technical, and other general tasks. Secretary interprets the law and other regulations, performs other tasks determined by the Statute, the Law, other regulations, and general acts of the Faculty. The Secretary participates in the work of the Faculty Council, without the right to vote. Its role in this organizational solution is for the session to take place under the rules of the Statute.

4.2 Smart Contracts

There are two types of smart contracts in the organization of electronic sessions. The first smart contract is the session itself. The Dean creates a new event called the Faculty Council Session. This smart contract consists of these attributes:

- full name of the Faculty
- ordinal number of the session
- date of the session
- time of the beginning of the session
- duration of the session
- number of proposed points
- list of all faculty council members.

The list of all faculty council members is just a list of public keys of individual members. Every employee will generate their own private key, with the help of the IT department, when he is employed, and there can be lots of lists. List of all employees, list of members of each department, list of members of some special committee, etc. It is proposed that the Secretary oversee creating and updating the lists.

This smart contract must have a method for creating session report and requires the same number of second type of smart contract as there is number of proposed points to vote on. So, the second type of smart contract is a proposed item. It was also created by the Dean and consists of these attributes:

- hash of session smart contract
- ordinal number
- title
- brief description
- link to documentation
- hash of documentation
- the question of accepting the item.

This smart contract must have a method for voting and returning the voting result. The properties title and description are part of every point. Each proposed point has the attributes of title and description, regardless of whether it is an electronic or regular session. The documentation, which usually consists of one or more documents, is published on the Faculty's website, and its hash is property of a smart contract to ensure traceability. In case the documentation contains more than one document, it is suggested that all documents be compressed into a.zip document and then the hash of that document be published in smart contract.

4.3 Blockchain Platform

There are various blockchain platforms that support smart contracts. Ethereum [11], Hyperledger [12], Tezos [13], Algorand [14], Solana [15] just to name a few. They are different by execution environment, smart contract language, permission type, consensus, etc. For this application, it is proposed to use the European Blockchain Services Infrastructure (EBSI) architecture, or in more detail CroBSI, the Croatian national blockchain infrastructure, that is subdomain of EBSI. It is important to state that EBSI is based on the Proof of Authority consensus.

The organization of holding and voting in the electronic session presented here is just one of the use cases and as such goes through four steps of development according to EBSI rules. EBSI documentation [16] gives these steps for Use Case Lifecycle:

1. Identification and selection
2. Design and development
3. Testing and piloting
4. Deployment in production.

By the time of drafting this article, there are four EBSI services:

- The Self-Sovereign Identity Use Case
- The Diploma Use Case
- The European Social Security Pass Use Case
- The Document traceability Use Case.

There are also plans to deploy more services like: Document Traceability, Trust Data Sharing, SME Financing and Asylum Process Management [17]. Illustration of EBSI layers interactions based on a simple blockchain transaction flow is presented in Fig. 3.

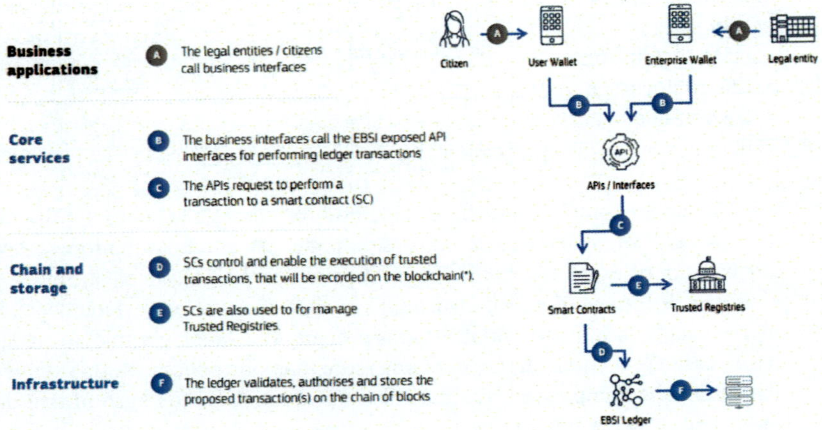

Fig. 3. EBSI layers interactions [16].

4.4 Voting Application

The application architecture itself is not the subject of this article. The application, by which Faculty Council members vote in sessions, must be created as a wallet library. Wallets are used for key management, signing blockchain transactions and reference implementations. Wallets can be installed on smart phones or other electronic devices (laptops, desktops, tablets, etc.), and they allow the user to access electronic sessions and vote at individual points. To activate the wallet user needs to enter the private key that he was given by a state administration body or after employment at the Faculty.

Current sessions will be visible in the wallet, on the day of the session, from the time set as the beginning, and voting on each item will be possible until the duration of the session expires. Within each proposed item, the member will be able to read all proposed item attributes, choose one of the offered answers ("FOR", "AGAINST" or "ABSTAINED") to the asked question, and submit the ballot. After selecting the answer and submitting the ballot, he will no longer be able to change his answer. After member casts vote on each item, the session will automatically close, and he will be able to see it under archive activities.

4.5 Session Report

After all members have voted on all proposed points or after the end of the session, the Secretary may request that the Session report be downloaded. That report must contain all attributes of the session smart contract, and for each smart contract of the proposed items must, in addition to all the attributes, fetch the results of the vote. It is understood that hash pointers to these blockchain transactions are seen in all parts of the report.

5 Conclusion

The current organization of voting in electronic sessions of the Faculty Council of FERIT has shortcomings in the organization and implementation. Using e-mail as a medium to deliver the agenda of the session as well as all the necessary attachments is not that as big of the problem like, voting on proposed points via e-mail. Agreeing or disagreeing with the proposed item is done by writhing a string of text. That string is prone to errors and is not entirely secure way of voting.

The proposed DLT technology is tested and offers unambiguity of each session and each proposed item. Using smart contracts that will automatically save the data in blockchain and create a report is the best way to prove voting transparency. Using smart contracts offer additional options, such as automatic cancellation of the session if 50% of the council members did not vote. Also, after the session expires or after all members have voted the session could be automatically closed. It offers the possibility of verifying the authenticity of published documents that are attached to the proposed items and so on.

Although the advantages of this method of voting are clear, there are certain disadvantages. Voting by e-mail requires only a username and password to login. Because email is used for a number of other purposes, people rarely lose or forget their username and/or password. In the case of blockchain voting, a person is much more likely to lose private key. On the other hand, remembering is not even an option. Thus, losing the key equals losing the ability to vote within a given time period.

Using the same organization, smart contracts can be made for voting at other sessions held at FERIT, such as voting by members of each department, and voting by members of various committees. In these cases, it would be necessary to make new lists of members and give the Presidents, of department and committee, the power to create sessions.

References

1. Epstein, J.: Are all types of internet voting unsafe? IEEE Secur. Priv. **11** (2013)
2. D. Uršić, Izrada akata u visokom obrazovanju s posebnim osvrtom na statut. In: Proceedings of 6th International Conference "Vallis Aurea" Focus on: Research & Innovation Požega: Veleučilište u Požegi; DAAAM International Vienna (2018). https://urn.nsk.hr/urn:nbn:hr:112:478386
3. Statut Fakultetskog Vijeća Fakulteta Elektrotehnike, Računarstva I Informacijskih Tehnologija Osijek, Osijek (2021) [Online]. https://docs.google.com/viewerng/viewer?url=https://www.ferit.unios.hr/2021/./dokumenti/429/Statut+FERIT+pro%C4%8Di%C5%A1%C4%87eni+tekst-o%C5%BEujak+2022.pdf. Last Visited: 10 June 2022
4. Poslovnik O Radu Fakultetskog Vijeća Fakulteta Elektrotehnike, Računarstva I Informacijskih Tehnologija Osijek, Osijek (2019) [Online]. https://docs.google.com/viewerng/viewer?url=https://www.ferit.unios.hr/2021/./dokumenti/70/Poslovnik+o+radu+Fakultetskog+vije%C4%87a+FERIT-a.pdf. Last Visited: 10 June 2022

5. Li, J., Kassem, M.: Applications of distributed ledger technology (DLT) and Blockchain-enabled smart contracts in construction. Autom. Constr. **132**, 103955 (2021)
6. Williams, S.P.: Blockchain: the next everything. Scribner (2019)
7. Park, S., Specter, M., Narula, N., Rivest, R.L.: Going from bad to worse: from Internet voting to blockchain voting. J. Cybersecur. **7**(1), tyaa025 (2021). https://doi.org/10.1093/cybsec/tyaa025
8. Jafar, U., Aziz, M.J.A., Shukur, Z.: Blockchain for electronic voting system—review and open research challenges. Sensors **21**, 5874 (2021). https://doi.org/10.3390/s21175874
9. Bulut, R., Kantarcı, A., Keskin, S., Bahtiyar, Ş.: Blockchain-based electronic voting system for elections in Turkey. In: 2019 4th International Conference on Computer Science and Engineering (UBMK), pp. 183–188 (2019). https://doi.org/10.1109/UBMK.2019.8907102
10. Abuidris, Y.: Secure large-scale e-voting system based on blockchain contract using a hybrid consensus model combined with sharding. ETRI J. **43** (2020). https://doi.org/10.4218/etrij.2019-0362
11. Ethereum Development Documentation [Online]: https://ethereum.org/en/developers/docs/. Last Visited: 10 June 2022
12. A Blockchain Platform for the Enterprise [Online]: https://hyperledger-fabric.readthedocs.io/en/release-2.2/. Last Visited: 10 June 2022
13. Welcome to the Tezos Developer Documentation! [Online]: https://tezos.gitlab.io/. Last Visited: 10 June 2022
14. Algorand Developer Docs [Online]: https://developer.algorand.org/docs/. Last Visited: 10 June 2022
15. Developer Resources [Online]. https://solana.com/developers. Last Visited: 10 June 2022
16. CEF Digital—EBSI Architecture, explained. Final draft 10 June 2021
17. Discover EBSI's USE CASES [Online]: https://ec.europa.eu/digital-building-blocks/wikis/display/EBSI/Use+cases. Last Visited 10 June 2022

Mobile Application for Keeping Records of Class Attendance

Petar Nenadić(✉), Damir Blažević, and Tomislav Keser

Faculty of Electrical Engineering, Computer Science and Information Technology, Osijek, Croatia
petar.nenadic@student.ferit.hr

Abstract. Recording attendance in the class meeting session is often time-consuming, prone to human error(s) and frequent recording of incorrect data. Also, tracking attendance per student can be a tedious process since it involves some sort of manual tracking/counting by pen and paper. This method consumes additional time and data may be prone to error from time to time. To avoid these problems, this paper describes a mobile application for student attendance. This application is mainly designed for students and professors who maintain attendance. Using this application, students can mark their attendance with mobile devices, furthermore, application allows the professors to keep track of student's attendance and store data for further retrieval. With advent of smartphones and tablets, which are very convenient to use, this process can be fully automated using the right technology. The tools used for application development are Kotlin and XML.

Keywords: Android · Mobile · Application · Records · Attendance · Student · Professor

1 Introduction

In current scenario, tracking attendance in the class session is the basic task of the subject handlers, since attendance marking can regulate class attendance and check the number of students in the class. Management and maintenance of student information is a key task of every institution. The task of marking attendance has traditionally been carried out manually with a diary. Later, this task is performed by the desktop applications. A desktop application is a standalone application installed on specific desktop or laptop computer and tasks can only be performed with that specific desktop system.

The main disadvantage of this system is that computer systems are not portable, so they cannot be kept anywhere to perform the task such as attendance recording. The entered attendance can be seen on an individual system if the desktop is connected to the network. Another method for reporting presence is the web application. In this method, attendance data is uploaded to the server over the Internet and users, such as students and professors, can view attendance through a web browser using any device, desktop, laptop or handheld mobile device. This system is only active when the internet connection is on because the data is not updated by the local database.

D. Blažević et al. (Eds.): OTO 2022, LNNS 592, pp. 37–44, 2023.
https://doi.org/10.1007/978-3-031-21429-5_4

These limitations of traditional systems are overcome by mobile applications. The mobile application allows users to install this application on their mobile devices. The professor can update the attendance data and update the data on the server. In order to reduce physical work and achieve greater efficiency in managing student information, this paper presents a mobile application based on Kotlin for easier and more efficient management of class attendance. The proposed application can store student information in a server database and can be retrieved by mobile devices. Through this system, professors can easily record attendance.

2 Mobile Application for Keeping Records of Class Attendance

This section explains the mobile application environment and the attendance management system mobile application. Figure 1 shows the architecture of the mobile application environment. An application for the attendance management system was developed and implemented in a cloud server such as Firebase. This application is also installed on mobile devices of users such as students and professors to access class attendance data. Users are divided into two groups: students and professors. Students can see the attendance that has been uploaded to Firebase. Professors can update student's attendance via mobile device.

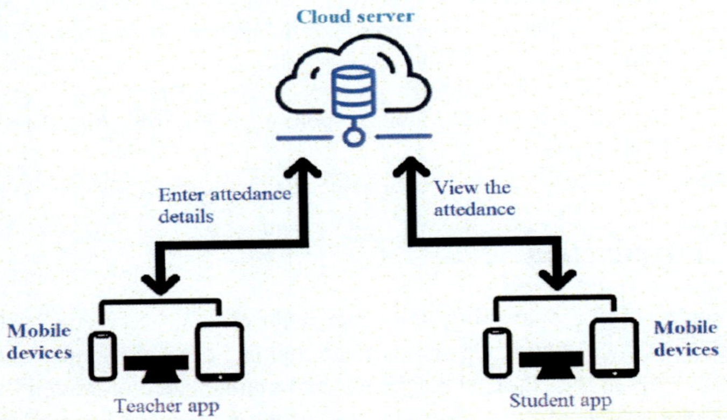

Fig. 1. Mobile application architecture for an attendance management environment

Figure 2 shows a schematic diagram of a mobile application for an attendance management system. This mobile application consists of two login modules namely staff login and student login. The student login enables students to view the data on class attendance that students receive. The professor login allows professors to update and review student attendance. Attendance data is stored in the database.

Figure 3 shows use case diagram for students and professors. Student can update his/her information and password, view attendance for each course and scan QR code from professor's mobile device. The professors can generate QR code for the students to take the attendance, check attendance data and see underrated students.

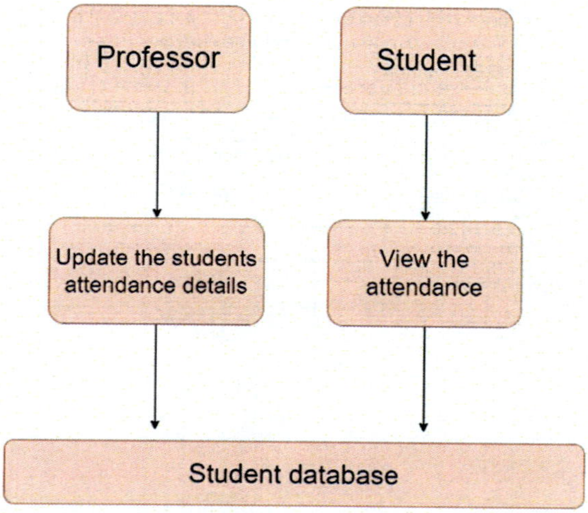

Fig. 2. Schematic diagram for attendance management system

3 Used Technologies

This mobile application is developed using Android Studio and Kotlin programming language. The application is implemented with the specification of the computer system with Windows 10 operating system, 8 GB of RAM memory and 250 GB SSD disk with Intel Pentium B960 CPU @ 2.20 GHz. The following procedure carried out for Kotlin based mobile application development.

3.1 Kotlin

Kotlin [6] is currently on of the most modern programming languages even though it started its development in 2011. It is a language that is constantly evolving and is constantly being improved with new features and functionalities. Since it enjoys a lot of support from Google, further support for Kotlin is expected, which is why more and more developers are using it instead of Java [7].

Kotlin can also be used to develop Web and server applications [8]. Although it is currently impossible to use it to develop applications for the iOS operating system, there is also a version of Kotlin Native that should allow developers to create applications for Android and iOS at the same time due to the support for the development of cross-platform applications (Fig. 4).

3.2 XML

XML [9] is a language used to mark-up data and documents. The idea behind this language is to create a language that can be read and understood even by the non-programmers. It is designed to store data in text form and frame it with tags so that they know what kind of data it is.

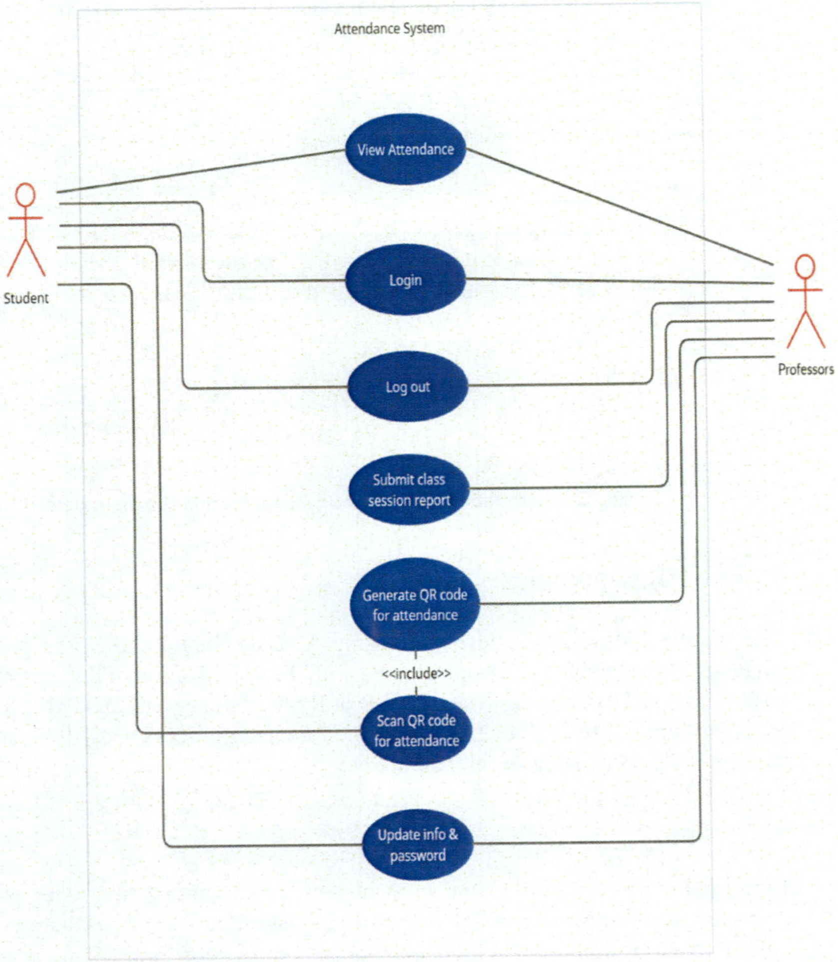

Fig. 3. Use case diagram for attendance

XML emphasizes usability over the Internet, simplicity and generality, it has strong support through Unicode for different languages. Although XML is intended primarily for use in documents, it can also be used for other purposes, such as data exchange, separation of data from presentation, data storage and increased data availability. XML is standardized language, which was taken care of by the World Wide Web Consortium [10].

For this project, XML is used to create layouts for application, input forms, buttons, images, creating backgrounds, etc. Example code is shown on Fig. 5.

```kotlin
class AboutActivity : AppCompatActivity() {
    override fun onCreate(savedInstanceState: Bundle?) {
        super.onCreate(savedInstanceState)
        setContentView(R.layout.activity_about)

        supportActionBar!!.title = "O nama"
        supportActionBar!!.setDisplayHomeAsUpEnabled(true)
        supportActionBar!!.setDisplayShowHomeEnabled(true)
    }

    override fun onOptionsItemSelected(item: MenuItem): Boolean {
        when (item!!.itemId) {
            android.R.id.home -> {
                finish()
            }
        }
        return super.onOptionsItemSelected(item)
    }
}
```

Fig. 4. Example of code written in Kotlin programming language

```xml
<ImageView
        android:layout_width="match_parent"
        android:layout_height="match_parent"
        android:src="@drawable/side_nav_bar" />
```

Fig. 5. Example for inserting image into layout of application

3.3 Firebase

Firebase [5] is platform of the company of the same name, which is used for the development of mobile and web applications, which can be used to quickly create application and increase the base of engaged users. The platform includes several well-integrated features that can be combined, and has analytics tools, mobile backends and tools to develop and maximize application success.

Firebase is used for this application because it enables real-time data storage and that data is available anywhere in the world. Figure 6 shows example table from Firebase.

4 Results and Discussion

This section presents sample code and results of the developed application. Figure 7 shows an example of the mobile application code for the student attendance monitoring system. Figure 8 shows the professor's login screen (a) and main screen (b).

Figure 8 shows screenshots from professor's application. Professors use their ID and password to login and gain access to information about students' attendance.

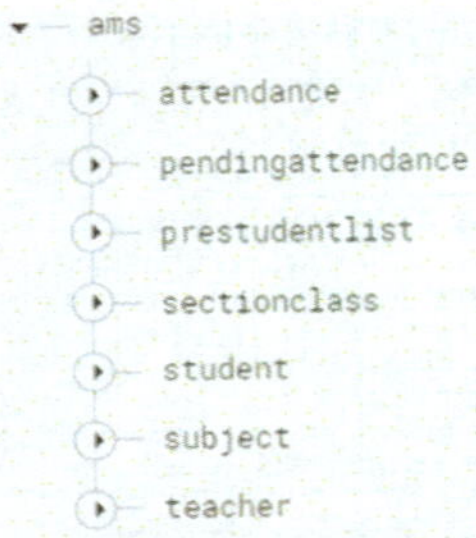

Fig. 6. Table from Firebase real-time database

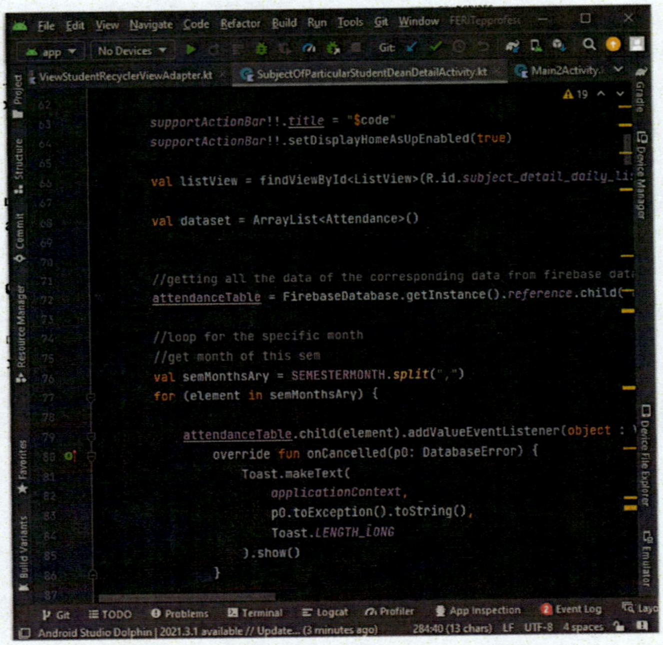

Fig. 7. Mobile application code example for attendance management system [16]

After professor is logged in, he/she can generate QR code which allows students to scan it with their mobile devices and confirm their attendance. Figure 9(a) shows application for professors with generated QR code and application for students for scanning QR code (b).

After student scans QR code from professor's applications, his attendance is recorded in database (Fig. 10). Those records are available to both professors and students.

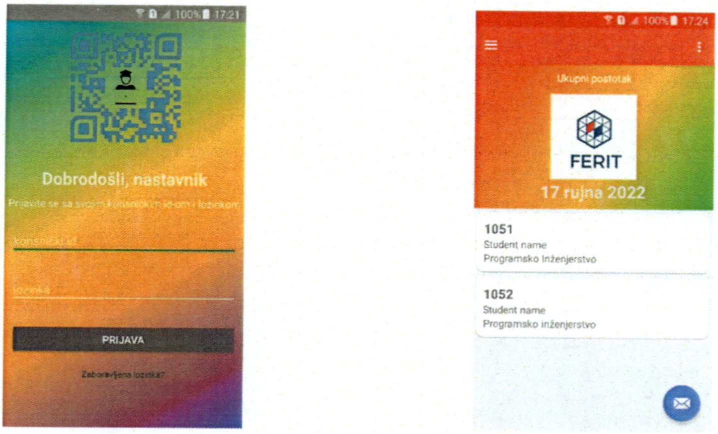

(a) (b)

Fig. 8. Professor login screen (a) and main screen (b)

(a) (b)

Fig. 9. (a) QR code from professor's application for attendance submission, (b) scanning QR code with camera from student's application

5 Conclusion

This paper presents a mobile application for the attendance management system. This system is enabled with two applications, one for students and one for professors. Registering attendance can be very time-consuming process, which request manual steps and is error prone. This application solves all the downfalls of registering attendance manually. It allows student to scan QR code from professor's mobile phone and register presence, which not only saves time but is also error free. All data is stored in Firebase database and is available anytime and everywhere. Also, professor can look at attendance of students whenever he/she wants and see which student has minimum attendance.

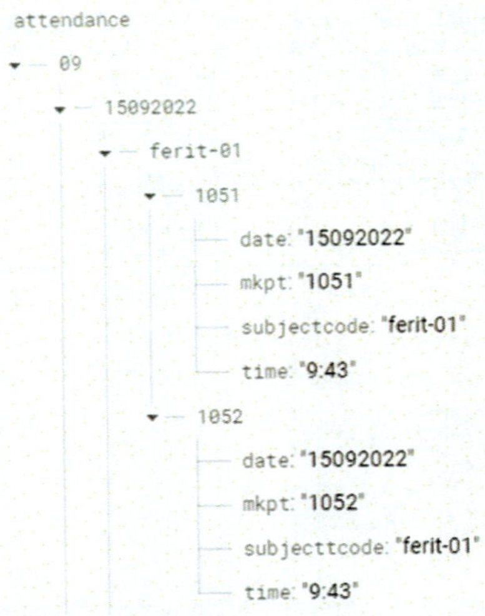

Fig. 10. Example of data stored in real-time database [5]

References

1. Android Developers: developer.android.com
2. Android Developers Blog: http://android-developers.blogspot.com/
3. Darwin, I.F.: Android Cookbook. O'Reilly Media, Inc. (2012)
4. Stack Overflow: http://stackoverflow.com
5. Firebase: firebase.google.com
6. Kotlin: kotlinlang.org
7. Lackner, P.: Kotlin Newbie to pro.https://www.youtube.com/playlist?list=PLQkwcJG4YTCRSQikwhtoApYs9ij_Hc5Z9
8. Griffiths, D., Griffith, D.: Head First Kotlin, 1st edn. O'Reilly Media (2019)
9. Horton, J.: Android Programming with Kotlin for Beginners. Packt Publishing (2019)
10. Isakova, S., Jemerov, D.: Kotlin in Action, 1st edn. Manning (2017)
11. Kotlin Programming: The Big Nerd Ranch Guide, Big Nerd Ranch Guides, 1st edn (2018)
12. Cafer, F.: Attendance Taker (2015). https://play.google.com/store/apps/details?id=com.ferid.app.classroom
13. Nirvan, R.: Attendance Register (2014). https://play.google.com/store/apps/details?id=com.rudra.attendanceRegister
14. Jibble: Free Time Clock App. Jibble Ltd (2020). https://play.google.com/store/apps/details?id=io.jibble.androidclient
15. Brar, A.: Virtual Attendance (2014). https://play.google.com/store/apps/details?id=deftsoft.virtual.attendance
16. Wifi Attendance, Redbytes Software Pvt Ltd (2016). https://play.google.com/store/apps/details?id=com.edsys.wifiattendance

Comparison Functionalities of HTTP and MQTT Protocols

Tomislav Kušević(✉), Damir Blažević, and Tomislav Keser

Faculty of Electrical Engineering, Computer Science and Information Technology Osijek, Osijek 31000, Republic of Croatia
tkusevic@gmail.com

Abstract. The mission and goal of this paper was to compare functionalities for both MQTT and HTTP protocol on the same example to see the advantages and disadvantages of sending and receiving data from server/device. Used tools are Android Studio, Kotlin, NestJS, TypeScript, PostgreSQL and in implementation two protocols were used: HTTP and MQTT protocols to determine advantages and disadvantages of the protocols, several examples are given where a particular protocol could be used in the development of mobile/IoT applications. While talking about MQTT and HTTP protocols, HTTP is one of the globally most used protocols for communication on the World Wide Web (Internet), when MQTT is for sure most used for connecting more devices in the networking called IoT (Internet of Things). Main goal is to show that MQTT can be also used in the mobile application, for example as notification service (server is notifying mobile devices).

Keywords: MQTT · HTTP · Android · Development · IoT · Protocols

1 Introduction

A few years ago, it was noticed an increasing number of devices that require connection to the Internet and access to Internet data. Main functionality is to connect multiple devices into one unit. There is also an increasing number of Internet protocols (HTTP, MQTT, etc.) that are used to connect these devices into one unit and manipulate data between them.

The main goal of the paper is to compare the functionality of the HTTP and MQTT protocols. It can be said that both protocols use the TCP/IP connection underneath, due to these similarities it is possible to compare them and get a clearer overview of the purpose of each protocol.

To successfully compare those two protocols, the paper will be based on practical work that will give us a more detailed overview of each of the protocols and how they work. Furthermore, this paper will be divided into two parts—the first part will be the practical part, previously described, while the second part is the theoretical part in which each protocol, its purpose and application will be described.

D. Blažević et al. (Eds.): OTO 2022, LNNS 592, pp. 45–56, 2023.
https://doi.org/10.1007/978-3-031-21429-5_5

1.1 Aim of the Paper

The goal of this paper is to create an application that will use the HTTP and MQTT protocol to connect and communicate with a remote server or other devices. This approach will help to have a clearer insight into each protocol, advantages, and disadvantages for both, based on which we will discern the purpose and expediency of the mentioned protocols. The application itself is quite simple, which gives us space and time to deal with the protocols themselves and their functionalities.

This application can add, edit, delete and read notes from a remote server. In addition, all devices will be notified when a change occurs in the database, so it will simulate the operation of an IoT system that mostly uses the MQTT protocol and try to find its usage in the development of mobile applications.

2 Used Technologies

In this part of the paper, the technologies that we used will be described will allow us to gain insight into the purpose and major features of the HTTP and MQTT protocols. It is well known that both HTTP and MQTT protocols are on the application layer of the TCP/IP protocol. To understand what the application layer is and what other layers are there, in the next chapter it will be explained what the TCP/IP protocol is and how it works.

2.1 General About the TCP/IP Protocol

TCP Protocol

TCP (Transmission Control Protocol) as a protocol on the application layer is responsible for creating the communication channels that make the network. It also determines how the message will be split into packets before they are sent to their destination using the network. Once the packets have successfully reached their destination, then the next step is to assemble the packets in the correct order so that they are understood and validated.

IP Protocol

IP (Internet Protocol) is a network protocol whose purpose is to route and address each packet so that each of them arrives at its destination. It is based on IP addresses written in packet headers. The most important data recorded in the IP structure are the IP address of the sender and the IP address of the destination [1].

TCP/IP Protocol

Using TCP and IP, TCP/IP (Transmission Control Protocol/Internet Protocol) was created. The first version that was more widely used was IPv4, which is still the dominant protocol used as communication over the Internet, later (in 2006) IPv6 was created—a more advanced version of the protocol that is gaining momentum. So, TCP/IP has been the standard for many years that provides computers and devices with remote communication over the Internet using packets [2, 3].

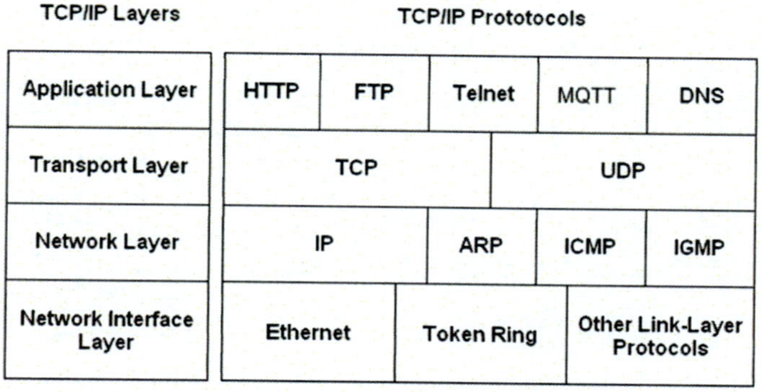

Fig. 1. TCP/IP architecture layers.

Each message is also sent in smaller packages that are usually sent by different routes to the same destination, where they are assembled and allow recipient to read the message at the destination. To understand how all of it is possible, it is best to describe each of them individually. TCP/IP consists of 4 layers: application, transport, network, and network access layers, which are shown in more detail in Fig. 1.

2.2 HTTP Protocol

The HTTP (Hypertext Transfer Protocol) protocol is an application protocol, which is the basis of the World Wide Web (Internet) data communication and is one of the most widely used protocols in the world. Currently, the HTTP/1.1 and HTTP/2.0 protocol versions are in widespread use (there is also HTTP/0.9). HTTP protocol communication works according to the request-response principle within the client-server architecture (Fig. 2). The most common clients are Internet browsers, mobile applications, client applications [5, 6].

Communication using the HTTP protocol takes place so that the client sends an HTTP request for a resource (e.g., an HTML page or a search operation) and receives an HTTP response from the server.

The request contains information about the identification of the resource (URL of the resource), the version of the protocol (e.g., HTTP/1.1), the HTTP method (GET, POST, PUT, DELETE, HEAD, OPTIONS or TRACE), request headers that define communication parameters (e.g., "Accept Language, Server: Apache 1.1") and request body with data to be sent to the server (e.g. user data in JSON format).

The response contains information about the protocol version, a status code (e.g., 200 OK) that provides the requesting client with information about the request processing process, response headers (e.g., Content-Type: text/html) and the response body with data that the server returns to the client (e.g., the corresponding JSON response).

Status Codes

- 100—Information while processing the request is still in progress
- 200/201—Success

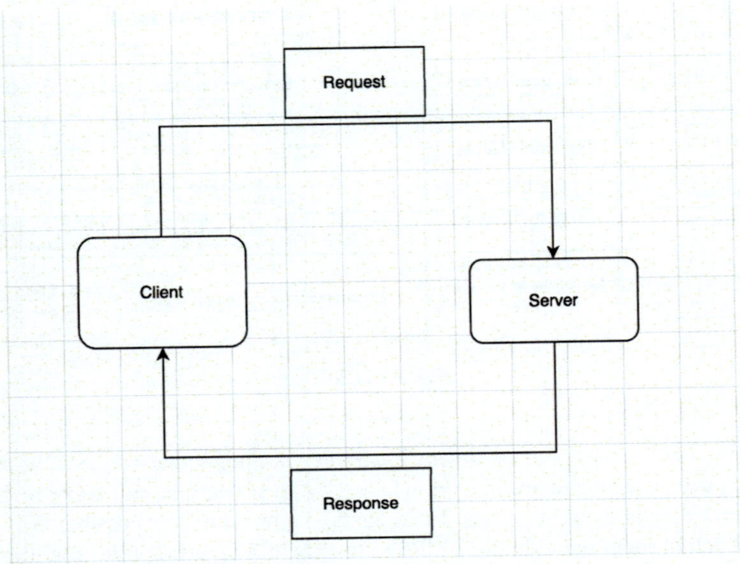

Fig. 2. HTTP Request-Response flow.

- 300—Redirect
- 400—Client error
- 500—Server error.

2.3 MQTT Protocol

General about the MQTT protocol:

MQTT (English Message Queuing Telemetry Transport) is one of the oldest M2M (English Machine to Machine) communication protocols. Andy Stanford-Clark from "IBM" and Arlen Nipper from "Arcom Control Systems Ltd" developed MQTT in 1999 [7, 8]. The main characteristic of the MQTT protocol is the lightweight messages, which are extremely suitable for IoT systems. MQTT consists of a client and an intermediary that communicate with each other so that the client publishes messages, while the server forwards the same messages to clients that have subscribed to that topic. Each message is posted to an address known as a topic. Clients can subscribe to multiple topics and receive every message published for each topic [7–10].

The most essential functions of the MQTT protocol:

Most often, when we talk about the MQTT protocol, we mention some of the basic functions that are necessary for the client (in this case, the device) to successfully receive data from the source (in this case, the broker), namely: connect, disconnect, publish, subscribe, unsubscribe [11].

2.4 Kotlin

Kotlin is a cross-platform programming language that is concise, secure, interoperable, and quite easy to use with other tools. It is a statically typed programming language that runs on the Java virtual machine and can also be translated into JavaScript source code, which makes it comprehensible to many developers who develop applications in programming languages: Java, JavaScript or TypeScript. Developed by the company "JetBrains", known for the development of powerful integrated development interfaces for the Java programming language called "IntelliJ IDEA". Today, Kotlin is the preferred programming language for developing Android applications—as confirmed by Google [12].

2.5 NestJS

NestJS is a platform (hereafter English framework) based on the Node.js platform, which is intended for use with the TypeScript (JavaScript) language, which aims to develop scalable applications on the server side as quickly and easily as possible. Applications developed on the NestJS platform are based on communication packages such as Express or Fastify. Nest is a new Node.js platform that not only imitates but also corrects the shortcomings of previous Node.js versions. When you start a new NodeJS project, NestJS is a much better choice than ExpressJS, because the intended project architecture is already defined in advance with a few simple components (controllers, modules, and services). This makes splitting applications into microservices simpler [13].

2.6 PostgreSQL

PostgreSQL is an open-source database. It is an object-relational database management system that stores data in rows, with columns as different data attributes. It allows you to store, process, and retrieve data safely. It was developed by a worldwide team of volunteers.

According to the DB-Engines PostgreSQL is currently ranked 4th in popularity amongst hundreds of databases worldwide [14]. PostgreSQL has built a formidable reputation for its proven architecture, reliability, data integrity, robust feature set, extensibility, and the commitment of the open-source community behind the software to consistently deliver efficient and innovative solutions. PostgreSQL runs on all major operating systems. It is no surprise that PostgreSQL has become the choice of many because of its features. We used PostgreSQL as a database where we store notes and users. Below is a list of tables within the database used in the project (Fig. 3).

Core features:

- PostgreSQL is compatible with Linux, Windows and macOS operating systems.
- PostgreSQL is highly secure, robust, and reliable. PostgreSQL supports multiple programming interfaces such as Java, C, C++, and Python.
- PostgreSQL supports various data types such as integer, string, and boolean. It also supports structured data types such as date/time, array, range. It can also work with documents like JSON and XML.
- PostgreSQL supports Multiversion Concurrency Control (MVCC) [15].

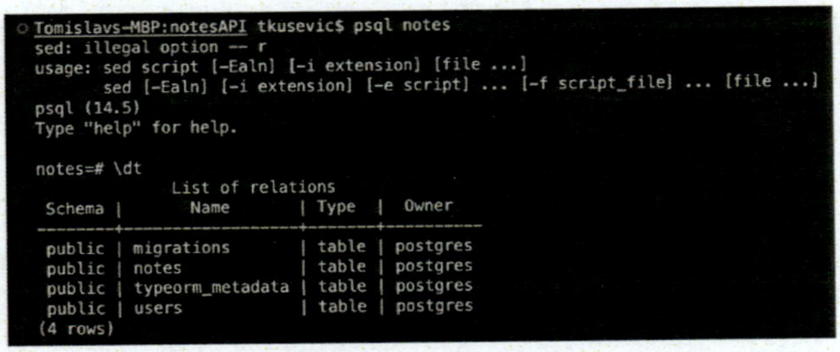

```
Tomislavs-MBP:notesAPI tkusevic$ psql notes
sed: illegal option -- r
usage: sed script [-Ealn] [-i extension] [file ...]
       sed [-Ealn] [-i extension] [-e script] ... [-f script_file] ... [file ...]
psql (14.5)
Type "help" for help.

notes=# \dt
            List of relations
 Schema |       Name        | Type  |  Owner
--------+-------------------+-------+----------
 public | migrations        | table | postgres
 public | notes             | table | postgres
 public | typeorm_metadata  | table | postgres
 public | users             | table | postgres
(4 rows)
```

Fig. 3. PostgreSQL list of tables used in project.

3 Implementation of the Program Solution

The implementation of the software solution can be divided into 2 basic parts:

- Flow chart development
- Backend development
- Development of a mobile (Android) application that communicates with the server via HTTP protocol (Request & Response) and MQTT protocol (Publish & Subscribe).

3.1 Flow Chart Development

In the diagram (Fig. 4), clients (Android applications) connect to the server via the HTTP protocol with requests and wait for responses because of each request, and according to the MQTT Broker, after connecting, they have the option of pre-subscribing to a specific topic. Also, the server could connect to the MQTT Broker and with the "Publish" function, it could send a specific message to all clients.

3.2 Backend Development

For our application to function, it was necessary to initially create and develop prerequisites to save certain data in the database, clients (devices) have access to this data and can change, view, add and delete it. We used the previously mentioned PostgreSQL database as a database (local, later uploaded database on the server via the Heroku service). The access to the application itself is at the internet location: https://notes-mqtt.herokuapp.com/, and the documentation (for which we used an extension called Swagger) is at the internet location: https://notes-mqtt.herokuapp.com/api.

Data access is enabled on different routes and using different HTTP methods. All routes that can access the application are in the documentation and shown in the next picture (Fig. 5).

There are some of the most used routes that allow clients (devices) to access data from the server:

- POST/user/register—route used for user registration
- POST/auth/login—route used for user login

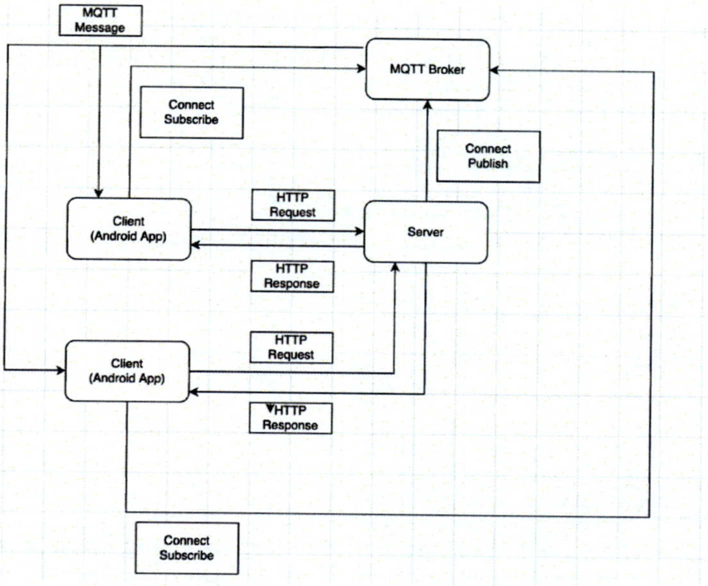

Fig. 4. Flow chart for the application solution.

Fig. 5. Swagger routes documentation.

- GET/auth/user—retrieving data of the currently logged in user
- GET/notes—get all notes

- GET/notes/:id—retrieving a specific note by identifier (id)
- DELETE/notes/:id—deletion of a specific note by identification mark (id)
- PATCH/notes/:id—edit a specific note by identifier (id).

3.3 Mobile (Android) Application Development

Functionality of the mobile application.

The developed Android application consists of:

- Splash screen
- Register screen
- Login screen
- Notes details screen
- Notes screen.

Splash screen

The functionality of the mobile application starts from the first initial window—the launch window (splash screen), where the memory in the device is checked to determine whether there is a logged-in user on the mobile device. If the user has already registered and logged in before, his identification number will be written into the memory, and the application will redirect him directly to the application, if the user has not previously logged in (or has deleted the application's internal memory), he will be redirected to the registration window.

Register screen

If the user does not have a registered account, this process is enabled on the register screen. The required registration parameters are e-mail and password. Besides the e-mail must be a valid e-mail, the password must consist of at least 6 characters to be valid. For registration, the server uses the JWT (JSON Web Token) package and the JWT strategy that enables the authentication process (registration and login). Also, after successful registration, the user will receive a token in the server's response with which he can access the rest of the call to the server (route).

Within each call, the client must send a token to get a valid response from the server, but also the server knows exactly which client wants access to the data and whether it has the right to the particular action it is trying to perform. The e-mail and password are saved in the database so that they can later validate a specific user during the login to the system.

Login screen

If the user has created an account within the application, then he can complete the application process on the application window. The registration process requires the e-mail and password of the previously created account in the registration window. If the login process is carried out successfully, the user is granted access to the application, more precisely to the notes window (list of notes).

Notes details screen

In addition to the list of notes, the user has the option of viewing all the details of a particular note. By clicking on a specific note, a new window will open showing all the details. When displaying the details, user see some of the following information: title of the note, text of the note, date the note was added, name of the author of the note. The user has the option to delete the note (if he is the author of it) by pressing the text "delete" in the upper right corner. In addition to deletion, the user also has the option to change the title and text of the note.

Notes screen

The main part of the application is the notes screen. Inside that window there are two tabs: the "All notes" tab and the "My notes" tab. In the notes screen there is a list of all notes in which user can see notes from all the users (Fig. 6 left), while in the "My notes" tab the user has an overview of the notes that he entered into the application (Fig. 6 right). Within the notes window, first the notes are retrieved from the server with an HTTP call and an MQTT connection is registered. The application is subscribed to a topic named "notes" which listens to all messages sent to the topic "notes"—this allows the user to retrieve data from the server again upon certain changes in the database (added or deleted notes) and to the list is refreshed.

Thus, user have the possibility of refreshing the list of data without the need for manual refresh. So there user can see the advantages of using the MQTT protocol within a mobile application. The user can delete his own notes directly from the list of notes. It is also possible to add a note by pressing the "+" sign in the lower part of the screen, after which a window appears in which you can add the title and text of the note.

4 Comparison of MQTT and HTTP Used in Project

With the development of this application, MQTT has its future in the development of mobile applications. As it was already stated, the HTTP protocol has been used for many years and in an exceptionally significant percentage for retrieving data from the Internet (servers). Although the MQTT protocol is mostly used in IoT systems, now they have an example of using this form of protocol in the world of mobile applications as well [16]. The real advantage of MQTT over HTTP occurs when we reuse the single connection for sending multiple messages in which the average response per message converges to around 40 ms.

When choosing MQTT over HTTP, it is so important to reuse the same connection as much as possible. If connections are set up and torn down frequently just to send individual messages, the efficiency gains are not significant compared to HTTP.

Greatest efficiency gains can be achieved through MQTT's increase in information density for each payload message. Most straightforward approach is to reduce the payload size where more data can be transmitted in each payload, which can be achieved through choosing proper compression and package methods based on the type of the data being generated [17]. For an easier comparison, follow three tables can show certain information and data about the use of both protocols (Tables 1 and 2).

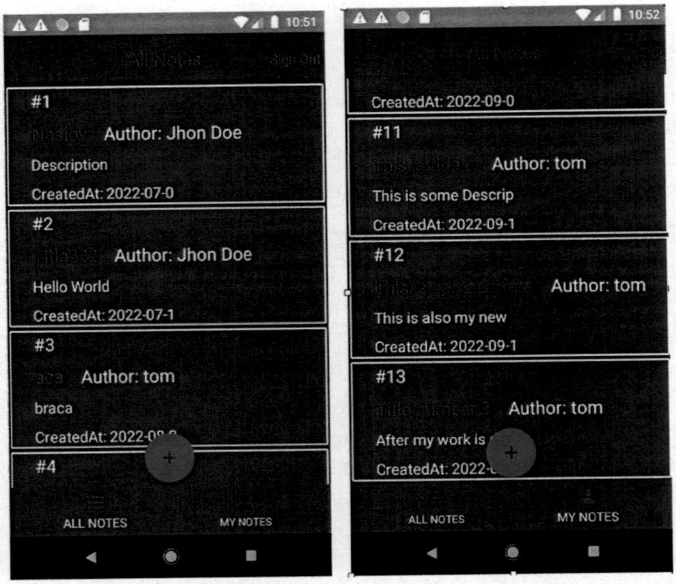

Fig. 6. Display of two tabs in Notes screen.

Table 1. Specification table for HTTP and MQTT.

	MQTT	HTTP
Name	MQ telemetry transport	Hyper text transfer protocol
Arhitecture	Publish/subscribe	Request/response
Aim	Topic	URL
Based on the protocol	TCP/IP	TCP/IP
Security	TLS + username/password	TLS + username/password
Type of connection	Connection status known	Connection status not known
Sending process	Async, event based	Sync
Queue	Broker can remember requests and informations	Based on implementation
Header size	2 bytes (binary)	min. 8 byte (text)
Message size	max. 256 MB	–
Message type	No type specified (mostly binary)	Text (Binary-Base64)
Message distribution	One-to-Many	One-to-One
Reliability	0—fire and forget 1—at least once 2—at most once	Based on implementation

Table 2. Comparison of memory usage in HTTP and MQTT functions.

Function name	MQTT (B)	HTTP (B)
Connect	5572	2261
Disconnect	376 (optional)	0
Sending message	388	3285
For 1 message	6336	5546
For 10 messages	9829	55046
For 100 messages	44,748	554,600

Table 3. Response time comparison for HTTP and MQTT.

No. of messages	MQTT—response time (ms) (QoS 1)	HTTP response time (ms)
1	113	289
100	47	289
1000	43	289

5 Conclusion

By comparing the HTTP and MQTT protocols on a simple application, there are advantages, disadvantages and intended operation of the protocol [15–17]. HTTP protocol is mostly used when a large amount of data should be sent (e.g. large JSON) to save certain data in the database. Every time client open an HTTP call, receive data, the connection is closed.On the other hand, the MQTT protocol is an excellent "light" protocol that allows us to connect two devices (subscribe to a topic) at once, and their communication is possible without closing it. If a device is subscribed to the topic, it will receive messages continuously.

In addition to two-way communication (a device can receive messages on a topic, but also send messages on the same topic), MQTT also allows us to communicate with multiple devices. Let us say there are 2 devices in the house that constantly send changed data (e.g., thermostat) that measure temperature and humidity—such a device can send data to a specific topic to all other devices that are connected to it. With that alone, it is necessary to send a message to the topic once, and all other subscribed devices will receive that message (the so-called broadcasting Data).

MQTT can also be used as a protocol for sending notifications to devices: Consider that there is an application that deals with writing and editing documents/notes. If two or more people were to edit the same document, a race-condition could occur over the data where one user would save the document after changes, and the other user would not catch those changes because he opened the document before editing and overrode the previous user's changes—in that situation MQTT can be of help to us. Each user, when opening the document, would subscribe to the topic, in case of a change to the document—all users would receive a notification that the document has been changed and to withdraw the last changes. Thus, race-condition is not occurring anymore. There is also some greater implementation of the MQTT protocol in the IoT industry (device communication), but also in application development—some form of the notifications.

References

1. What is the Internet Protocol? https://www.cloudflare.com/learning/network-layer/internet-protocol/. Last accessed 24 Sept 2022
2. Tobey, P.: IPv4 and Education: The Basics (2022). https://circleid.com/posts/20220607-ipv4-and-education-the-basics. Last accessed 24 Sept 2022
3. Ames, I.: TCP/IP Protocol Suite (2019). https://medium.com/software-engineering-roundup/tcp-ip-protocol-suite-ec7ed4888d5d. Last accessed 24 Sept 2022
4. GeeksForGeeks: Transport Layer Responsibilities. https://www.geeksforgeeks.org/transport-layer-responsibilities/. Last accessed 24 Sept 2022
5. Grigorik, I.: High Performance Browser Networking. O'Reilly Media (2013)
6. Fielding, R., Reschke, J.: Hypertext Transfer Protocol (HTTP/1.1): Message Syntax and Routing, RFC 7230. Internet Engineering Task Force (IETF) (2014)
7. Egli, P.: MQTT—Message Queueing Telemetry Transport Introduction to MQTT, A Protocol for M2M and IoT Applications (2017)
8. Šušnja, M.: Primjena mqtt protokola u internetu stvari, Specijalistički diplomski stručni, Sveučilište u Splitu, Sveučilišni odjel za stručne studije. Split (2020). https://urn.nsk.hr/urn:nbn:hr:228:047974
9. Bensakhria, A.: IoT—Communicating with Devices: Introduction to MQTT and HTTP (2020)
10. MQTT vs. HTTP: Which One Is the Best for IoT? https://medium.com/mqtt-buddy/mqtt-vs-http-which-one-is-the-best-for-iot-c868169b3105. Last accessed 24 Sept 2022
11. MQTT Beginner's Guide? https://www.u-blox.com/en/blogs/insights/mqtt-beginners-guide. Last accessed 24 Sept 2022
12. Heller, M.: What is Kotlin? The Java Alternative Explained. https://www.infoworld.com/article/3224868/what-is-kotlin-the-java-alternative-explained.html. Last accessed 23 Sept 2022
13. An Introduction to MQTT with NestJS Framework. https://myas92.medium.com/an-introduction-to-mqtt-with-nestjs-framework-a69ec55483f0. Last accessed 24 Sept 2022
14. PostgreSQL Tutorial. https://www.simplilearn.com/tutorials/sql-tutorial/postgresql-tutorial. Last accessed 25 Sept 2022
15. What is PostgreSQL and Why Do Enterprise Developers and Start-ups Love It? https://canonical.com/blog/what-is-postgresql. Last accessed 25 Sept 2022
16. MQTT Vs. HTTP for IoT. https://www.hivemq.com/blog/mqtt-vs-http-protocols-in-iot-iiot/. Last accessed 24 Sept 2022
17. HTTP vs. MQTT: A Tale of Two IoT Protocols. https://cloud.google.com/blog/products/iot-devices/http-vs-mqtt-a-tale-of-two-iot-protocols. Last accessed 25 Sept 2022

Influence of Wind Power Plant on P-V and Q-V Curves in Transmission System

Filip Pisonic, Kresimir Fekete(✉), Ruzica Kljajic, and Predrag Maric

Faculty of Electrical Engineering, Computer Science and Information Technology Osijek, University of Osijek, Kneza Trpimira 2B, Osijek, Croatia
{kresimir.fekete,ruzica.kljajic,predrag.maric}@ferit.hr

Abstract. Recently, an increasing number of wind power plants (WPP) or wind farms have been connected to transmission systems around the world. In this paper, the influence of wind farms on P-V and Q-V curves, which are related to voltage stability, is investigated. For the purpose of the aforementioned research, the IEEE 24 bus Reliability Test System (RTS) was modeled in the PowerWorld Simulator program. The model is then used for the calculation of P-V and Q-V curves for different scenarios of WPP location and size. For every scenario voltage profile (voltage magnitude for each of 24 buses in the system) is calculated as well as P-V and Q-V curves. The results are compared with the Base Case (i.e. the case in which WPP is not connected to the system). Also, results for the scenarios are compared and conclusions are made.

Keywords: P-V curve · Q-V curve · Wind power plant · Voltage · Transmission

1 Introduction

The large increase in the connection of wind power plants (WPP) to the transmission system rise the question of how such power plants will affect the voltage stability of the system. Recently, many research studies on the mentioned topic are available in the scientific literature. The literature [1] analyzes the voltage stability (P-V and Q-V curves) of a transmission system with a connected charging station for electric vehicles and a WPP in which different types of wind turbines are implemented. In the literature [2], the authors examine the voltage stability of the transmission system at the point of connection of the battery charger for electric vehicles and the distributed energy source (renewable energy source—WPP and solar power plant). For the benchmark of the transmission system. IEEE 14 bus system [3] is used. The influence of wind power plants on voltage and transient stability has been investigated in the literature [4]. The test was performed on a power system with 9 buses modeled in GE Positive Sequence Load Flow Analysis (PSLF) software [5]. Characteristics of different wind turbine generators used in modern WPP are presented in [6]. Some types of wind turbine are suitable for voltage regulation and thus can positively affect the voltage stability of the system. Evaluation of P-V and Q-V curves in inverter based (mainly WPPs) power system is presented in [7]. IEEE 9 bus test system is used as a model of transmission system [8].

D. Blažević et al. (Eds.): OTO 2022, LNNS 592, pp. 57–68, 2023.
https://doi.org/10.1007/978-3-031-21429-5_6

In this paper IEEE 24 bus test system [9] is used for investigation of WPP influence on the P-V and Q-V curves of the transmission system. The test system is modeled in PowerWorld Simulator [10] which is used for calculation of the P-V and Q-V curves as well. The WPP is connected to different location (defined by the different simulation scenarios) and bus voltage, active power losses and the P-V and Q-V curves are calculated respectively. Obtained results are finally compared and discussed.

The rest of the paper is organized as follows: used model and all the input data are presented in Sect. 2. Simulation scenarios and calculation results are given in Sect. 3. Comparison of the results is presented in Sect. 4 and conclusion is given in the last section.

2 Network Model and Input Data

2.1 Network Model

The IEEE 24 bus test system [9] is chosen as a representative example of transmission system. Although the original test system is made for testing various reliability analysis methods, it is also used for various voltage analysis [11]. For the purpose of the calculations that are made, the IEEE 24 bus network is modeled in PowerWorld Simulator. The single line diagram of the modeled system is depicted in Fig. 1.

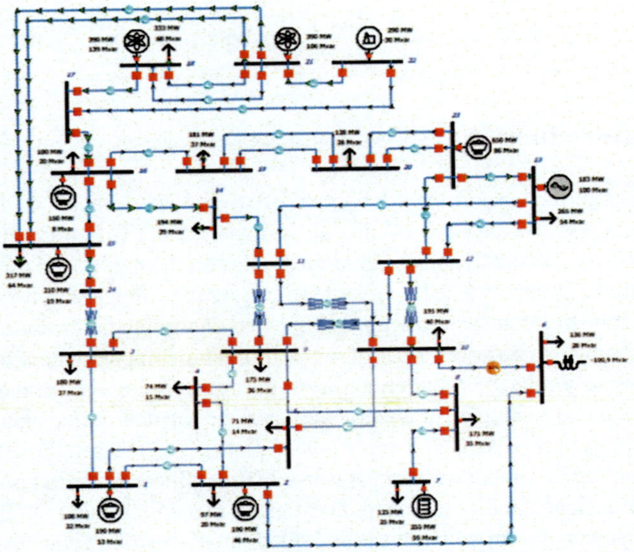

Fig. 1. PowerWorld single line diagram of the modeled IEEE 24 bus test system.

Table 1. Generators data.

Bus number	Output active power [MW]	Voltage regulator set point [p.u.]	Active power limit [MW]	Reactive power limit [MVAr]
1	190	1.035	192	−50 + 80
2	190	1.035	192	−50 + 80
7	255	1.025	300	0 + 180
13	n/a	1.020	n/a	n/a
15	210	1.014	215	−50 + 110
16	150	1.017	155	−50 + 80
18	390	1.050	400	−50 + 200
21	390	1.050	400	−50 + 200
22	290	1.050	300	−60 + 96
23	650	1.050	660	−125 + 310

2.2 Input Data

Set points for generators in the system (output active power, voltage regulation set point as well as active and reactive power limits) are shown in Table 1.

Bus 13 is selected as slack bus of the system and thus the data for output active power as well as active and reactive power limits are not applicable. The data for active and reactive consumption are listed in Table 2. The total active power consumption in the system is 2850 MW.

Table 2. Bus consumption data.

Bus number	Consumption active power [MW]	Consumption reactive power [MVAr]	Bus number	Consumption active power [MW]	Consumption reactive power [MVAr]
1	108	22	13	265	54
2	97	20	14	194	39
3	180	37	15	317	64
4	74	15	16	100	20
5	71	14	18	333	68
6	136	28	19	181	37
7	125	25	20	128	26
8	171	35			
9	175	36			
10	195	40			

3 Simulation Scenarios and Results

3.1 Simulation Scenarios

In order to investigate the influence of the WPP on the P-V and Q-V curves, five (5) simulation scenarios are defined as follows:

Scenario 1—base case scenario. There are no WPP integrated and in this scenario bus voltages are observed. The input data for generator production and load consumption are mentioned in Tables 1 and 2.

Scenario 2—200 MW WPP is integrated on buses 3, 4, 5 and 6. WPP participates in voltage regulation. In this scenario the best location for WPP integration is determined based on obtained Q-V curves.

Scenario 3—200 MW WPP is integrated on a bus 3 and two cases are analyzed. In the first case there is no reactive power regulation and in the second case WPP participates in reactive power regulation. PV and QV curves and bus voltages are given for both cases.

Scenario 4—WPP is integrated on a bus 3 and in first case installed power is 600 MW. There is no voltage regulation and bus voltages are observed and compared with Scenario 1. In second case, WPP power increases to 1045 MW and WPP participates in reactive power regulation. PV curves, QV curves and bus voltages are given for both cases.

3.2 Results

Power flow calculation is performed and results are obtained for each scenario. In base case scenario, operating point is changed in comparison to original IEEE 24 bus. As a result of decreased generator power, bus voltages are in range between 0.78 and 1.05 p.u.

Scenario 1 results. Bus voltages for scenario 1 are given in Fig. 2. If voltage limits are set within ± 10%, it can be noticed that some buses have voltages lower than desired limit. Lowest voltage is on bus 2, 0.79843 p.u. and the buses 18, 21, 22 and 23 have the highest voltage (1.05 p.u.).

In addition to huge voltage drops, there is also increased current in cables resulting in higher losses. Total losses in this case are 87.5 MW.

Scenario 2 results. In this scenario, the best location for WPP integration is determined based on bus voltages after integration. As shown in Fig. 3, voltages are higher on all buses after 200 MW additional power is integrated in system.

After results are obtained, set of results for bus 3 is taken and analyzed for various WPP power and ability to maintain voltage stability.

Scenario 3 results. First location for implementation of 200 MW WPP is bus 3. WPP does not participate in voltage regulation. After WPP is integrated, bus voltages increase since additional power is added to the grid, Fig. 4. The biggest increase is noticed on busses near place of WPP integration. WPP increases voltages near its location. Power flows have changed in a manner that total system losses decrease to 82.6 MW as well as cabel overload.

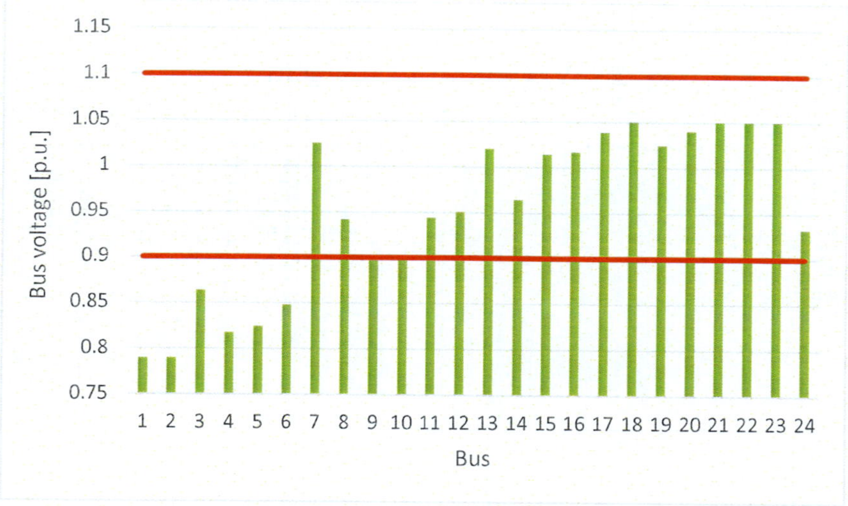

Fig. 2. Bus voltages, scenario 1.

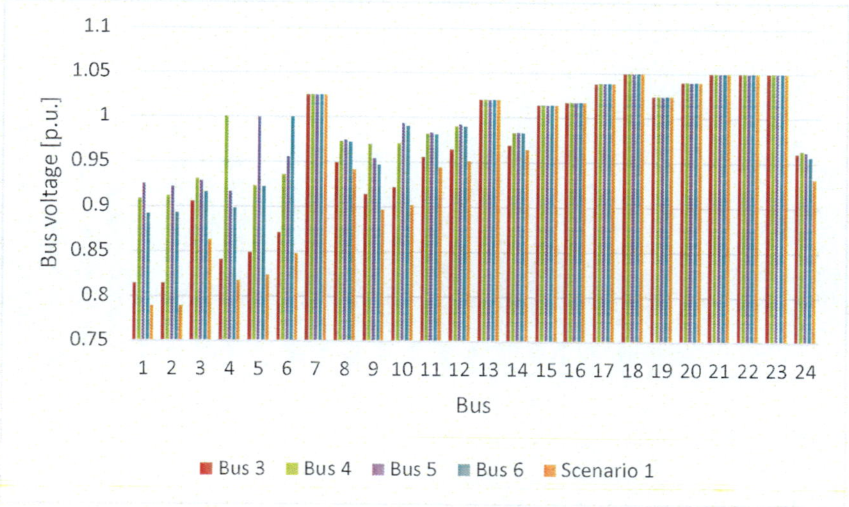

Fig. 3. Bus voltages, scenario 2.

In second case, WPP can participate in voltage regulation and it is integrated on bus 3. Voltage on bus 3 is set to 1 p.u. Fig. 5 shows that, in comparison to Scenario 1, voltages are higher on all buses as WPP increases voltage on all neighbouring buses. Losses are 75.8 MW and cable overload (cabel connecting buses 6 and 10) is still present.

Scenario 4 results. In the forth scenario two cases are analyzed. In the Case 1, WPP has installed power of 600 MW and WPP does not participate in voltage regulation.

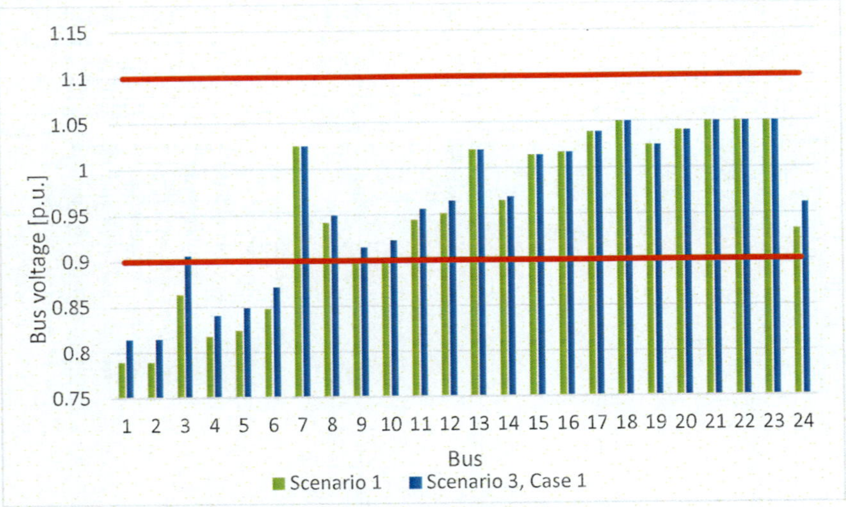

Fig. 4. Bus voltages, scenario 2 without voltage regulation.

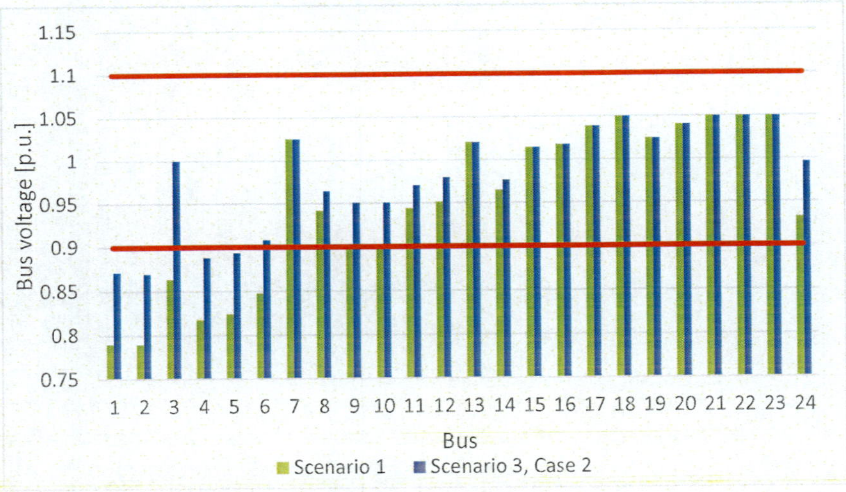

Fig. 5. Bus voltages, scenario 3 with voltage regulation.

Higher active power installed on a bus 3 results in higher bus voltages on the place of integration. Bus voltages are shown in Fig. 6. With higher active power, resulting power flows change and with higher power in system, losses increase. This results with lower bus voltages in comparison to scenario 1. This impact is mostly highlighted od busses 1 and 2.

System losses now have a value of 125.8 MW. Cable loadings have changed and cables that were overloaded in Scenario 1 (cable connecting buses 6 and 10) are now

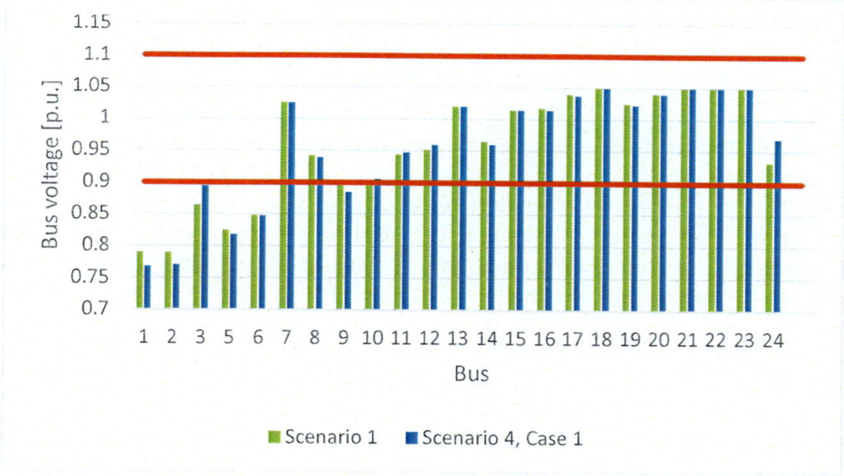

Fig. 6. Bus voltages, scenario 4 without voltage regulation.

under their loading limits. On the other hand, cables connecting busses 3–1 and 3–9 have increased loading, up to 55% of their rating. High WPP power can affect bus voltages in a negative way as it was shown in this scenario.

Results for Case 2 are given in Fig. 7. In this scenario, 1046 MW WPP is integrated on bus 3 and it can participate in voltage regulation. Local bus has higher voltage than in Scenario 1 but voltage decreases on other buses. In this case higher power is integrated on a bus in order to run WPP near its stability limits.

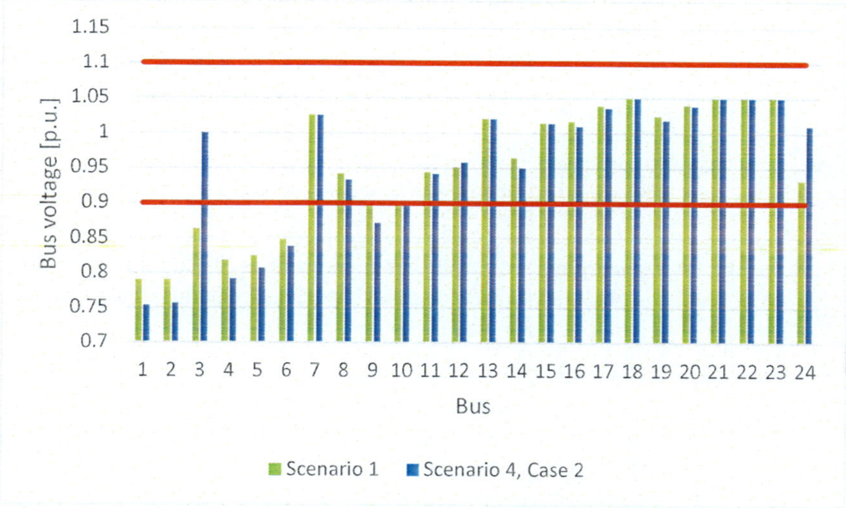

Fig. 7. Bus voltages, scenario 4 with voltage regulation.

Total losses have also increased and now have a value of 212.1 MW. Cabel between buses 6 and 10 is still overloaded as well as cables connecting buses 3–1 and 14–16. This change is a result of increased power flows due higher active power in system.

4 Discussion

P-V and Q-V curves are main tool in analyzing system stability. Some specific points in P-V curves are normal operating point, and voltage collapse point. In this paper, stability margin is set on 0.85 p.u. (red line). If voltage decreases under this value it is considered that voltage stability cannot be maintained.

Scenario 2—Q-V curves. Since WPP participates in voltage regulation, voltage is set to 1 p.u. Maximum active power that can be transferred is determined by WPP power. With increased power this margin also increases. On the other hand Q-V curves give insight in voltage stability in case that voltage is fixed (Fig. 8).

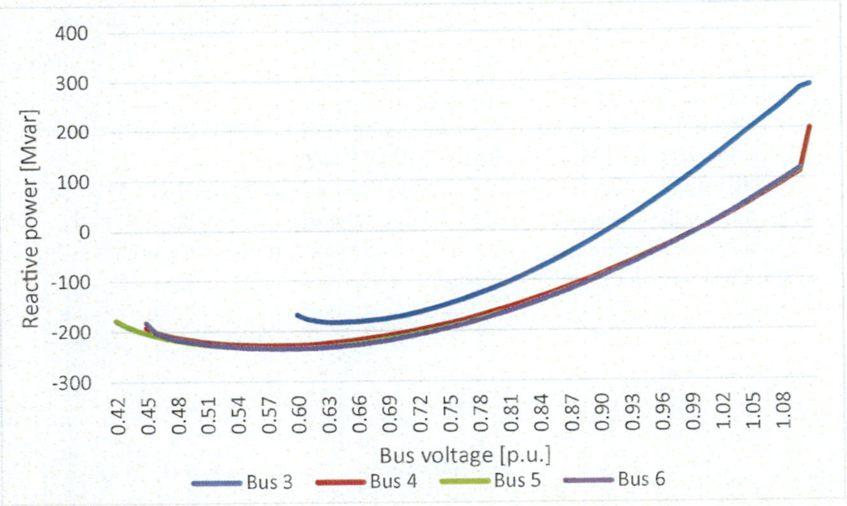

Fig. 8. Q-V curves, scenario 2.

The lowest margin is for WPP integrated on bus 3. Due this, the most interesting set of results is obtained for WPP on bus 3. Increasing active power results in lowering stability margin and it can even lead to instability.

Scenario 3—P-V and Q-V curves. P-V curve for both cases is shown in Fig. 9. In Case 1 initial voltage is 0.906 p.u. and active power increases in steps of 1 MW. As active power increases, voltage begins to decrease. Critical point is 0.85 p.u. and it is considered to be a point after which voltage stability cannot be maintained. As it can be seen, active power corresponding to knee point is 587 MW and after this point voltage increases rapidly. When active power reaches value of 601 MW, voltage decreases to 0.833 and after this voltage collapse occurs.

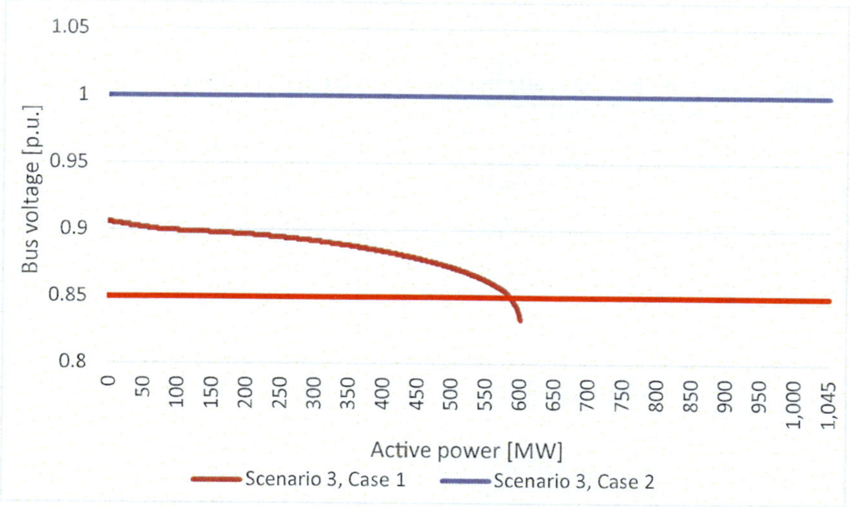

Fig. 9. P-V curves, scenario 2

In Case 2 WPP participates in voltage regulation. PV curve changes in comparison to previous scenarios and now it has fixed value of 1 p.u. No matter how much active power increases, voltage remains 1 p.u. Maximum power that can be transferred before voltage collapse is 1045 MW.

QV curve for given case is shown in Fig. 10. In Case 1, maximum loading before collapse is around 190 Mvar. System is stable as increase in reactive power results with increased voltage. System stability can also be confirmed by applying dQ/dV criteria as its derivative is greater than 0.

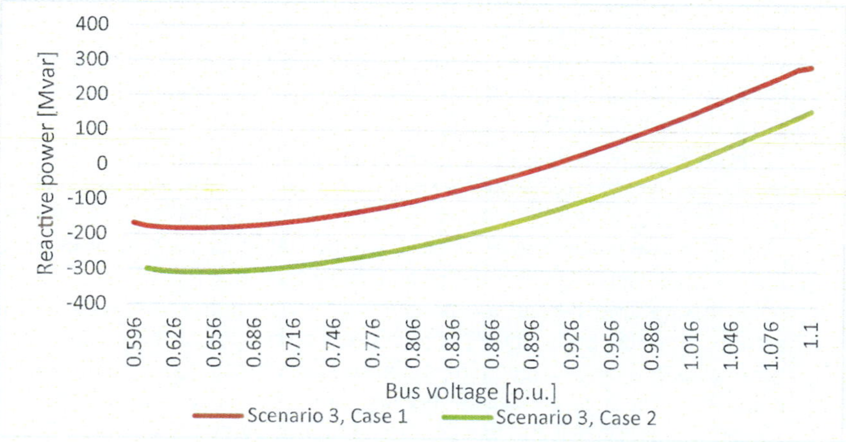

Fig. 10. Q-V curves, scenario 3.

QV curves also differ in Case 2. Reactive power margin with 300 Mvar is lower in comparison to Case 1. Voltage stability is enhanced in comparison to previous scenarios. Also, stability margins have increased in this case.

Scenario 4—P-V and Q-V curves. In Scenario 4, WPP works near its stability margin. The biggest change can be noticed when PV curves are observed. In previous Scenario, maximum power that can be transferred is 600 MW. With higher integrated power stability margin decreases and voltage collapse happens when active power increases to 320 MW (Fig. 11).

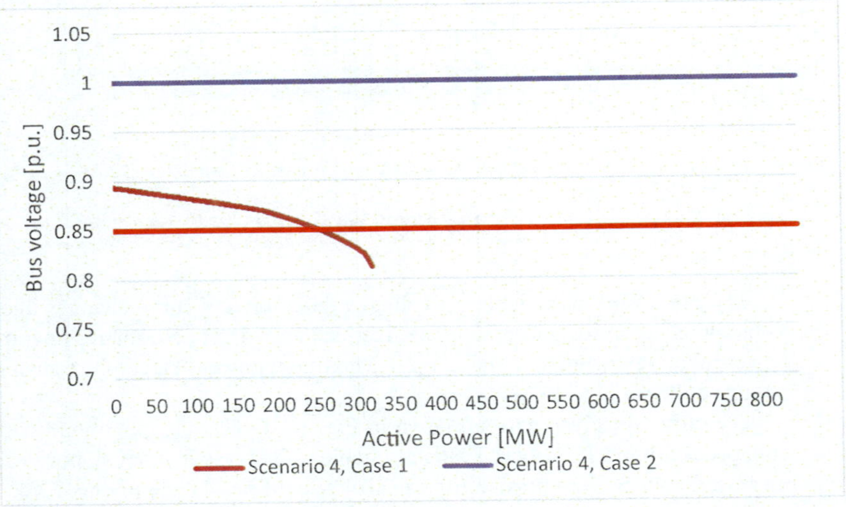

Fig. 11. P-V curve, scenario 4

Since WPP participates in voltage regulation, PV curves have a specific value of 1 p.u. WPP maintains this value on a given bus and maximum power that can be transferred is 840 MW. After this value, voltage collapse occurs. If regulation is not possible, maximum active power is 600 MW.

In comparison to Scenario 3, stability margins have decreased, Fig. 10. The minimal reactive power has decreased and stability margins are also smaller. System is stable in both scenarios, as $dQ/dV > 0$ and operating point is located to the right of the minimal point. Reactive power before collapse is now 100 Mvar. In both cases, reactive power limit is negative but higher value is preferable (Fig. 12).

In this scenario there are two possible equilibrium points. In order to determine which point is operating point dQ/dV criteria needs to be applied. First point (intersection of curve with 0 Mvar, voltage 0.86 p.u.) is not stable since $dQ/dV < 0$. Other intersection point (0 Mvar, 1 p.u.) is stable because $dQ/dV > 0$. In comparison to all previous scenarios, this scenario has the narrowest stability margins and only 60 Mvar are needed to collapse voltage.

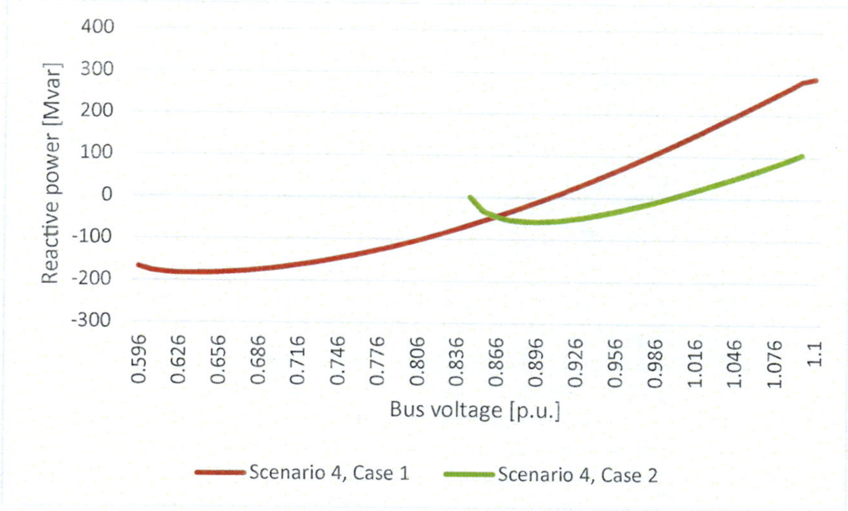

Fig. 12. Q-V curves, scenario 4.

5 Conclusion

Maintaining voltage stability is important task and various methods are used to achieve it. The transition to green energy has led to an increased integration of RES. The integration of RES can affect the stability of the system. Positive effect on voltage values and consequently reduced losses can be achieved by good selection of location and proper sizing of RES. WPP, as one of the most common RES, can adapt to any grid and enhance all aspects of stability. Installation close to the point of consumption can raise the voltage locally and reduce currents in the network, which will consequently reduce losses. High installed power can also have a negative effect, since power flows increase with higher active power, which results in greater voltage drops. This influence can also be considered from the stability point of view by studying the PV and QV curves. If WPP is integrated in the system, the voltage stability is improved. However, high integrated power leads to a breakdown in voltage stability. If RES is already integrated into the system, stability can be further improved by enabling voltage regulation. The voltage regulation is coupled with the reactive power of the generator, i.e. it is enough for the WPP to provide certain amounts of reactive power and thus affect the voltage magnitudes. When WPP participates in voltage regulation, the limits of voltage stability increase. Using WPP in voltage regulation is desirable, however, high active power with demand to participate in voltage regulation can become new source of instability. In order to comprise these requirements, detailed studies need to be performed for each given grid configuration.

References

1. Gumilar, L., Cahyani, D.E., Sholeh, M.: P–V and Q–V curve analysis of four types wind power plants at battery charging station. In: International Seminar on Application for Technology of Information and Communication (iSemantic) 2020, pp. 249–254. Semarang (2020)
2. Gumilar, L., Sholeh, M., Netanel Rumokoy, S., Monika, D.: Analysis voltage stability in the interconnection of battery charging station and renewable energy. In: 2nd International Conference on Cybernetics and Intelligent System (ICORIS), pp. 1–6. Manado (2020)
3. IEEE 14-Bus System. https://icseg.iti.illinois.edu/ieee-14-bus-system/. Last accessed 23 May 2022
4. Muljadi, E., Nguyen, T.B., Pai, M.A.: Impact of wind power plants on voltage and transient stability of power systems. In: IEEE Energy 2030 Conference, pp. 1–7. Atlanta (2008)
5. PSLF Simulation Engine. https://www.geenergyconsulting.com/practice-area/software-products/pslf-re-envisioned. Last accessed 23 May 2022
6. Camm, E.H., et al.: Characteristics of wind turbine generators for wind power plants. In: IEEE Power & Energy Society General Meeting, pp. 1–7. Calgary, Canada (2009)
7. Sewdien, V.N., Preece, R., Torres, J.L.R., van der Meijden, M.A.M.M.: Evaluation of PV and QV based voltage stability analyses in converter dominated power systems. In: IEEE PES Asia-Pacific Power and Energy Engineering Conference (APPEEC) 2018, pp. 161–165. Kota Kinabalu, Malaysia (2018)
8. IEEE 9-Bus System. https://icseg.iti.illinois.edu/wscc-9-bus-system/, Last accessed 23 May 2022
9. IEEE 24-Bus System. https://icseg.iti.illinois.edu/ieee-24-bus-system/. Last accessed 23 May 2022
10. PowerWorld Simulator Homepage. https://www.powerworld.com/. Last accessed 23 May 2022
11. Rahman, T.K.A., Jasmon, G.B.: A new technique for voltage stability analysis in a power system and improved loadflow algorithm for distribution network. In: International Conference on Energy Management and Power Delivery EMPD '95, pp. 714–719 (1995)

New Trends in Power Transformer Surveillance and Diagnostics in the Function of Power System Maintenance

Igor Provči, Robert Noskov(✉), and Ivica Petrović

Croatian Transmission System Operator, 10000 Zagreb, Croatia
robert.noskov@hops.hr

Abstract. The restructuring of the electricity market that occurred in the previous years has led to a change in the approach to the maintenance of equipment, stations and grid elements. With the liberalization of the market, profit becomes the most important aspect in power system exploitation, hence, in order to avoid costs, there is a reduction in preventive maintenance procedures for both transformers and other equipment. This inherently reduces the safety and reliability of the system. In order to maintain the level of safety and reliability of operation at a sufficient level, they are replaced by systems of occasional or continuous monitoring and periodic thermal imaging. The most effected grid elements are energy transformers which are coherently the most important and most expensive elements of the grid. Determining the availability of power transformers is important for the safe operation of the power system. Transformer monitoring and diagnostic systems are particularly interesting for more efficient resource management, increasing system reliability and safety while preventing unwanted consequences or grid faults. Monitoring systems have poor diagnostic properties and are, therefore, combined or supplemented with other methods or procedures. Additionally, new diagnostic methods and techniques are being investigated that could continuously monitor and diagnose the condition of transformers. Thereby, different thermal imaging procedures present an unavoidable factor in proper transformer diagnosis.

Keywords: Monitoring · Thermovision · Power transformer

1 Introduction

The market approach has forced transmission and distribution system operators to choose appropriate maintenance procedures for their equipment, stations and grid elements. Otherwise, unplanned breakdowns and power outages would cause unacceptable costs or other economic losses. The fault reclamation time consists of the time required to diagnose the fault and the time to repair the fault. The goal of all methods of preventive maintenance, various diagnostic methods and various periodic inspections, monitoring and thermal imaging procedures is to reduce the number and duration of faults, reduce the duration of troubleshooting and reduce overall economic costs and its consequences.

With preventive maintenance, certain pieces of equipment are replaced at set intervals. Today, preventive maintenance is the most common principle of maintenance. The

D. Blažević et al. (Eds.): OTO 2022, LNNS 592, pp. 69–83, 2023.
https://doi.org/10.1007/978-3-031-21429-5_7

issue is that it usually requires a planned decommissioning, but since this decommissioning can be planned at the optimal moment, this method minimizes the costs of decommissioning. Preventive weather maintenance is expensive because the elements are changed according to a predetermined schedule, although these elements may still be in good enough condition. On the other hand, some elements may need to be changed before the time set by the maintenance schedule.

Increased pressure on costs and new technological possibilities of diagnostics have led to the improvement of preventive maintenance. Additionally, maintenance is no longer time based but also based on the element to the condition. The concept of predictive maintenance—maintenance according to the condition was introduced. A good example of condition-based maintenance is transformer oil analysis. Previously, the oil was changed, regardless of its condition after each predetermined period of time. Testing and analyzing the oil can increase its service life, reduce costs due to its expensive nature. The transformer is in overhaul less often and environmental problems are avoided as old oil needs to be disposed of properly.

The principle of condition-based maintenance is professional equipment based diagnosis to which a subjective, experiential diagnosis is added for achieving better results i.e. maintenance costs and the number of downtimes are reduced. Therefore, in addition to objectify measurement results and computational procedures, it is extremely important to model the experience of diagnostic experts from their daily practice. This method will be increasingly used in the future as it can significantly reduce losses for plant owners and increase availability.

The concept of proactive maintenance involves acting on the cause of the problem. The concepts of predictive and proactive maintenance, which are superior to traditional reactive and preventive maintenance, especially in terms of cost reduction and avoidance of sudden station faults are inconceivable without the organization and implementation of quality technical station diagnostics. For these reasons, a department of technical diagnostics is organized within the maintenance sector.

By applying reliability-based maintenance, each element of the station has its own minimum safety scheduled maintenance, which contributes to the overall increase in safety, reliability, as well as maintenance costs reduction. This method also takes into account the consequences of faults and outages, on the environment, drive and safety of people [1].

Figure 1 shows a holistic approach to developing a successful maintenance strategy based on operational reliability.

Diagnostics in maintenance should determine the condition of individual parts of the system without dismantling them, and preferably without their downtime. The basis of these diagnostics is the measurement of the selected parameter and the given value. By comparing diagnostic parameters or quantities, with predefined permissible values of these quantities, a decision is made on the availability of an individual grid element and the possible need of repair or replacement of a distinct component.

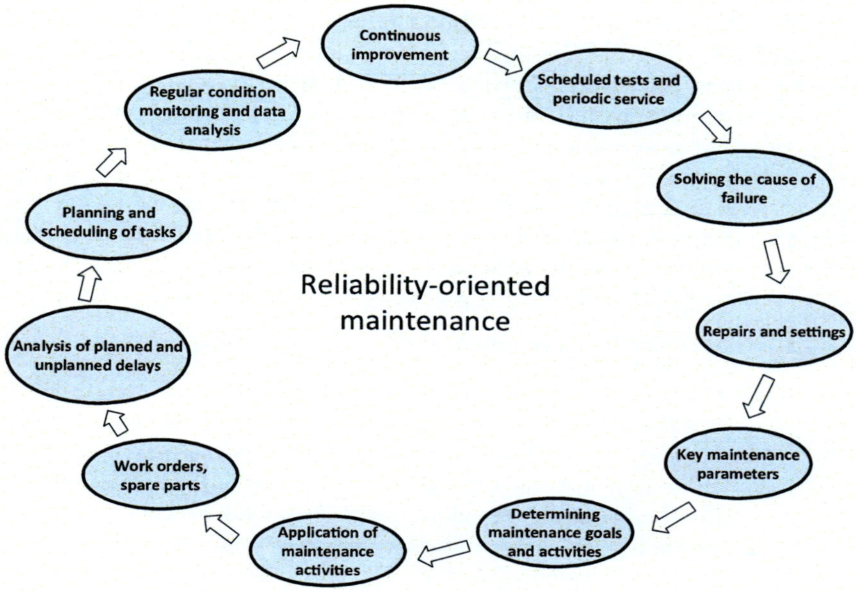

Fig. 1. Reliability-oriented maintenance.

2 Significance and Characteristics of Diagnostic Methods and Monitoring of Power Transformers

Diagnostic tests can be performed continuously or periodically at set times, and there is also the possibility of installing diagnostic devices in the station, which provide continuous diagnostic monitoring. Continuous monitoring is done constantly and is performed by an external device. Periodic inspection is performed at regular intervals, and can be performed by a device or a human [2–6].

The organization and implementation of diagnostic monitoring achieves the following beneficial effects:

- reduction of total costs,
- increasing the reliability of work,
- increase safety at work.

The cost reductions are manifested through:

- better planning of maintenance activities,
- reduction of consumption of spare parts,
- reduction of required spare parts stocks,
- lower energy consumption,
- avoiding accidents at the station.

It should be noted that the organization and implementation of diagnostic monitoring allows for better maintenance planning, as it provides a good insight into the condition of equipment. Therefore, before the overhaul of the transformer, its diagnostic inspection can be performed, which reveals its defects that need to be sorted out during the overhaul. According to some research, the application of technical diagnostic methods reduces the consumption of spare parts. The reason for reducing their consumption lies in the fact that the application of these methods provides insight into the true condition of the elements parts, which allows them to be replaced only when they are completely worn out without the threat that no suitable spare parts will be available in the critical moment.

2.1 Power Transformer Monitoring

As transformers are key assets in the power grid and industrial processes need to be constantly monitored to prevent an unexpected transformer failure that can be catastrophic. A major failure on a transformer is defined as an incident that takes the transformer out of service for seven or more days. To determine the risk of unexpected failure, it's important to consider the consequences and the probability of failure occurring. According to [6] about one out of every 200 transformers fails each year with major failures cost about $14,000 per MVA in property damage. Figure 2 shows major failure locations on power transformers.

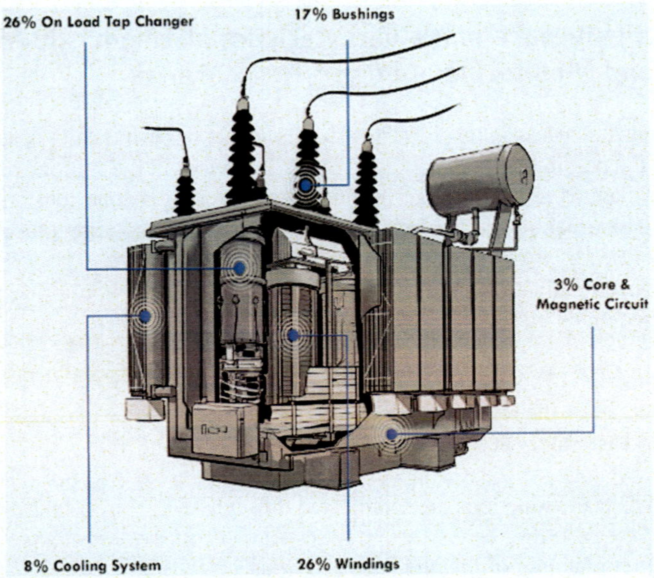

Fig. 2. Major failure locations on power transformers.

Transformer monitoring is continuous monitoring of the transformer condition, which basically includes measuring certain value and monitoring the condition of the transformer equipment, i.e. certain diagnostics of the condition of individual parts of the transformer (cooling system condition, sensor condition, etc.). Furthermore, it is possible to estimate certain parameters based on measurements and mathematical models, and to archive measured and estimated parameters. Monitoring of power transformers also contains a user interface for access to monitoring results (HMI—Human-Machine Interface).

Additionally, some monitoring systems have the ability to exchange data with other systems and devices in the station, such as SCADA system (Supervisory Control And Data Acquisition) which is shown in Fig. 3. Those systems have the ability to access the monitoring system from a remote location [3, 4], and transformer cooling system control options [5].

The following goals can be achieved by installing a monitoring system on the transformer:

- detection of faults as they occur and prevention or reduction of the consequences of those faults,
- constant insight into the operating conditions and condition of the transformer,
- condition-based maintenance,
- increase availability,
- optimization of transformer management,
- more detailed analysis of the causes of fault,
- increase in system security.

This method can detect changes in certain parameters that are measured using sensors or estimated using a specific mathematical model. Thus, for example, due to increased thermal or electrical stresses in the vicinity of the insulating material, it degrades its insulating properties, decomposes cellulose and oil and the forms various gases in the oil. Additionally, moisture and other decomposition products can occur.

Very reliable indicators in diagnosing mechanisms that can lead to transformer faults are:

- increase in oil and winding temperature,
- occurrence of partial discharges,
- change in capacitance and dielectric loss factors of conductors.

There are a wide range of indicators that can detect the symptoms of transformer failure by monitoring the cooling system, load tap changer, dissolved gas, bushing power factor and capacitance, partial discharge, oil levels, pressure, temperatures and more. Some of them will be described in the following text.

Analysis oil dissolved gases. Analysis of oil dissolved gases is one of the most reliable diagnostic methods introduced in the mid-1960s [8]. It is most often performed by periodically taking oil samples from transformers, typically once a year. In laboratory conditions by chromatographic analysis the concentrations of gases dissolved is then determined. These gases most often consist of: hydrogen, carbon monoxide, carbon dioxide, ethylene, ethane, methane, acetylene and oxygen. The development of sensors

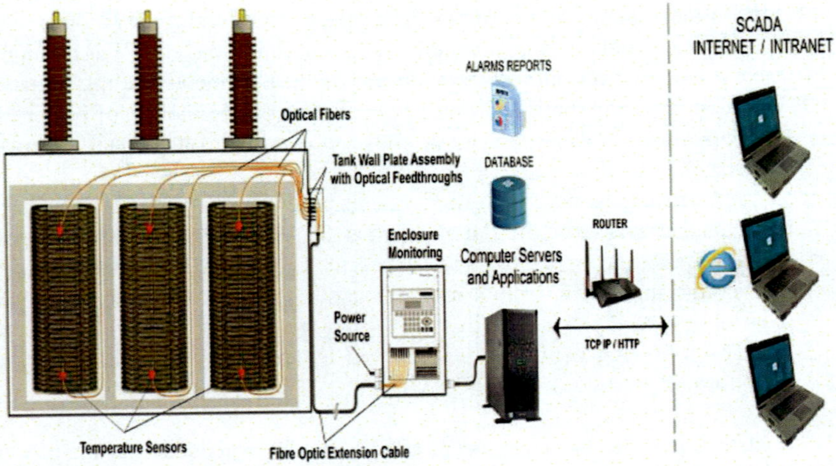

Fig. 3. Monitoring system integrated within power transformer and presented in SCADA system [7].

for on-line measurement of the concentration of gases made it possible to bridge the time between periodic tests of oil samples in the laboratory. These devices allow the measurement of a mixture of gases dissolved, and are mainly used to warn of the increased generation of gases in oil.

In addition to measuring oil dissolved gases, these sensors also allow the measurement of the increase in moisture dissolved, which is a product of degradation of paper insulation that reduces the dielectric strength of the insulation of the transformer.

Temperature Monitoring. By monitoring the temperature, it is possible to determine the phenomena of overheating in the transformer, to assess the efficiency of the cooling system of the transformer, and to assess the state of insulation of the transformer. The most important temperature, on which the aging of the insulation directly depends on, is the temperature of the hottest point of the winding.

The temperature of the hottest point of the winding can be measured directly by installing special optical thermometers that measure the temperature at one point or along the entire winding. Such models take into account the oil temperature, the load factor of the transformer and the inclusion of pumps and fans.

Voltage and current monitoring. Voltage monitoring is most often performed by measuring the voltage at the measuring connection of the conductor. In addition to voltage, it is also possible to measure changes in conductor capacity, which is a direct indicator of potential conductor failure. Conductors are exposed to high electrical and mechanical stresses, and the most common failure mechanisms are moisture penetration and partial discharges.

Current is most often measured using current measuring transformers. It is an extremely important monitoring parameter, because, in combination with temperature monitoring, it allows estimating the temperature of the hottest winding point, which results in the aging speed of paper insulation and estimating the remaining life, and allows transformer overload planning.

Partial discharge monitoring. Partial discharges occur as a result of an increase in voltages, insulation damage, moisture in the insulation, cavities in the solid insulation, loose metal parts, and gas bubbles in the oil. An increase in partial discharges in the transformer insulation is a sign of weakening the insulation properties of the material and as a result, insulation breach is possible.

Electrical and acoustic methods are available for the detection of partial discharges. Acoustic sensors are installed in the transformer or are mounted externally on the transformer boiler. They are more sensitive to external interference (rain, wind, loose vibrating parts, core noise, etc.) but they are easier to install on an old transformer. The biggest advantage of the acoustic method is the ability to locate partial discharges within the transformer itself using an adequate algorithm and good sensor placement.

If partial discharges occur in the oil, there will be an increase in the hydrogen concentration which can be detected by some of the gas detection sensors.

Tap changer monitoring. Most transformer faults are caused by the tap changer. Although these are generally faults with minor consequences, they reduce the reliability of the transformer, which is why monitoring of the tap changer is desirable. Faults of the tap changer are mainly mechanical and electrical in nature. To diagnose mechanical faults, the torque of the switch drive motor is usually monitored. By measuring the oil temperature in the control switch boiler and comparing it with the oil temperature in the transformer boiler, it is possible to determine whether there is an increase in oil temperature of the tap changer.

Cooling system monitoring. Generally, the states of the pumps and fans are most often monitored. If some fans or pumps are not turned on when they should be, there will be an increase in the temperature of the transformer. By monitoring the condition of the cooling system, it is also possible to better estimate the temperature of the hottest point of the winding.

2.2 Further Development of the Transformer Monitoring System

Today, monitoring systems are recognized as an important tool for more efficient transformer management as well as an important component of any power station. From the current point of view, two directions for further development of the monitoring system are predictable.

The first involves the further development of sensors and measurement methods in order to improve existing ones in both technological and economic terms as well as developing new ones. It is primarily expected to improve the existing methods for monitoring partial discharges in the form of reducing external interference, followed by oil gases and conductors. One of the most promising new methods is the monitoring of vibrations of the tap changer [9] as a tool for the detection of mechanical and electrical disturbances in operation, as well as the development of sensors for measuring the content of furfural in oil.

In parallel with the development of sensors, the development of automated data analysis collected by the monitoring system into clear and meaningful information (knowledge of the state of the transformer) that is presented to the user is expected. Namely, as the number of parameters that can be monitored by the monitoring system increases daily, an increase in the amount of data collected also increases. To analyze such a large amount of data, it is necessary to spend some time, and equally important, to have the knowledge to interpret the results. Precisely because of the limited human resources with specialist knowledge, users are looking for a way to enable more effective interpretation of monitoring results. One of the ways to achieve this is to reduce the number of monitored quantities, which in turn directly reduces the value and limits the functionality of the monitoring system.

The solution to this problem is the development of a system that will, as an upgrade of the monitoring system, enable automated processing of monitoring results and, as a result, will provide better diagnostics of transformer conditions. In the technical literature, which covers the field of observation, a number of papers on this topic have been published in recent years. These are various attempts to process the data collected by the monitoring system using some of the artificial intelligence techniques and as a result generate certain recommendations, warnings, alarms etc.

The most frequently mentioned techniques are neural networks, expert systems, fuzzy logic, etc. These techniques are most often used to interpret the results of measurements of oil dissolved gases [10, 11], and to calculate the hottest winding point [12]. The results obtained show how neural networks can be successfully implemented as a tool for the classification of certain transformer states (fault predictions) as well as for the estimation of certain parameters. Neural networks are a suitable tool when there is monitoring data in normal operation of the transformer as well as in the fault condition, and when the connections between the input parameters in the algorithm and the output parameters from the algorithm are complex. The main advantages are the ability to learn and the resistance to noise in the signal.

Main disadvantages are the need to have data available in normal operation and at the time of fault as well as poor convergence of results. The main feature of fuzzy logic is the ability to process fuzzy information, which often appears in diagnostics (e.g. the amount of gas in the oil is "large", so it is not very precise). Fuzzy logic provides the possibility of combining neural networks and expert systems and upgrading them [13].

It is expected that the best results in online diagnostics could be achieved by integrating various methods into a single diagnostic tool. Apart from the combination of methods, other observed quantities should be taken into account, not only the concentration of gases in oil, which is the most common case.

3 Use of Thermovision in Transformer Diagnostic Control

Thermography of infrared, IR or thermal radiation has a very distinct place in the maintenance of installations, equipment and stations. The infrared scanning method itself aims to test the thermal distribution on the external, visible surfaces of electrical equipment and parts of the station that are in operation, without the need to turn them off or put out of operation. This technique, using the so-called "thermogram" visualizes, otherwise invisible, thermal radiation emitted by bodies, so this method is used in practice to detect increased heating of electrical and mechanical components during normal operating condition without contact with the object under test [14].

It is basically a method, which is based on thermal comparison of different objects, the so-called "measurement with reference", and refers to the comparison of identical spots, in different phases (L1, L2, L3) of the same bay or system. This requires a systematic scan of all three phases (preferably simultaneously) to determine any deviation from the "normal" thermal image.

Special attention is paid to coupling and suspension equipment, connection points of switches, disconnectors, measuring transformers, cable heads and insulators, and as for insulated parts (transformers, low-oil switches, metal-shielded SF_6 plants) thermal distribution is controlled over the entire outer surface. The criterion for determining the condition (severity of the fault) of the equipment on which the heating was observed, i.e. the degree of urgency of the required intervention, is taken as ΔT, the temperature difference between the observed "warm" and the reference point. This criterion was determined on the basis of long-term statistical monitoring of results and a large number of thermal imaging tests. When assessing the degree of urgency of the intervention, the voltage level of the equipment being controlled and the operating load at the time of the test have to be taken into account. This is quite important, as for higher voltage levels (i.e. larger dimensions and larger masses) relatively small increases in temperature can indicate quite serious faults. Condition of electrical equipment on which an increase in temperature has been observed can be estimated on the basis of the following criteria with an assumed 100% load. Table 1 shows an estimation of the degree of urgency of the intervention based on the temperature difference.

Table 1. Estimation of the degree of urgency for intervention based on the temperature difference.

T increase	State	Recommended measures
$0 \leq \Delta T < 5$ °C	0	Repeat the thermotest in 6–12 months
5 °C $\leq \Delta T <$ 10 °C	1	Intervene as soon as possible
10 °C $\leq \Delta T <$ 35 °C	2	Intervene the first time the operation is stopped

This diagnostic method prevents serious malfunctions, i.e. indicates poor or inadequate thermal insulation. In this way, the number of unplanned power outages is reduced, and thus the total downtime due to fault. On the other hand, it is possible to perform maintenance activities more rationally, which, in addition to shortening the time required for overhaul, also contributes to the quality of the performed interventions by concentrating on the thermal imaging detected problems.

The direct effects are manifested in the acceleration of the process of diagnosing faults and checking the undertaken interventions, saving energy, protecting capital equipment as well as reducing insurance premiums. Furthermore, by maximizing the availability of equipment, in addition to confirming its reliability but also pointing out possible critical spots, the total operating time is increased. From the aspect of safety, thermal imaging can effectively contribute to the detection of defects in material construction or monitoring of high risk processes.

In electrical engineering, electrical equipment testing most often indicates problems caused by current-resistance relationships. In general, a "hot spot" in an electrical circuit occurs as a result of insufficiently tightened, oxidized or corroded connection, but also improper operation of the appliance itself. Figure 4 shows an example of a hot spot.

When creating a station maintenance program, it is recommended to introduce the so-called maintenance cycle by means of thermal imaging monitoring of the state of electric equipment, which is schematically shown in Fig. 5, and which can be adapted to almost any industrial environment.

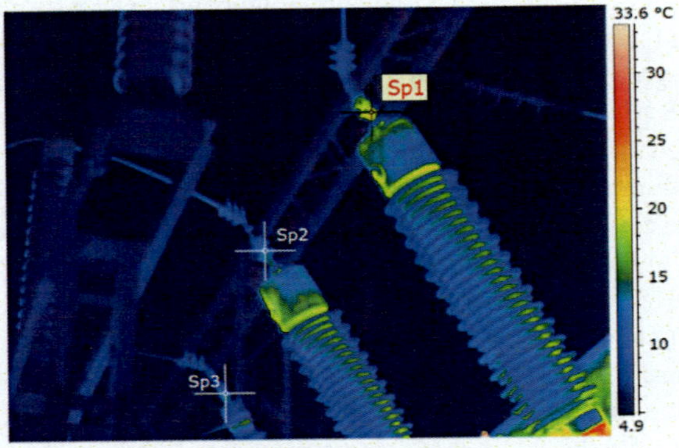

Fig. 4. "Hot spot" on the HV side of the power transformer.

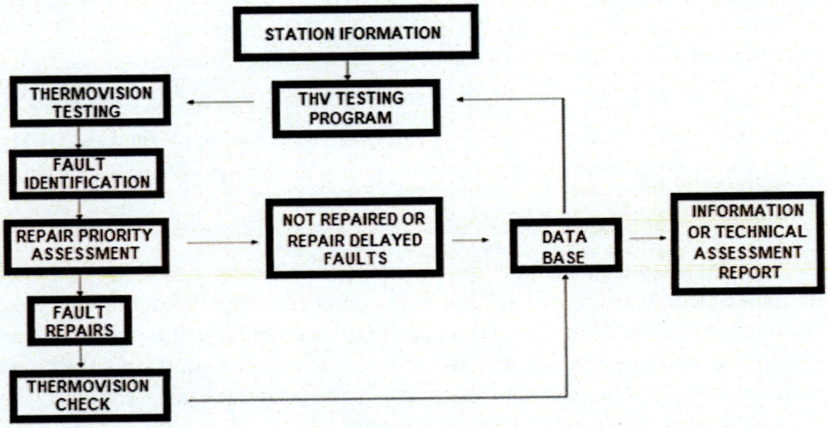

Fig. 5. Schematics of a thermovision state assessment maintenance cycle.

One of the more important actions that need to be performed is the classification of the equipment to be tested with regard to its strategic importance, possibility of replacement, operating age, safety standards and place of installation in the station. Based on this data, it is possible to create a thermal imaging program. After the test has been performed and the faults have been identified, it is necessary to start forming a list of priority repairs, based on the measured results, adopted criteria for assessing the condition, physical laws and empirical indicators.

The assigned priority determines the urgency of the intervention and refers to:

- emergencies requiring immediate shutdown and repair,
- situations where reparation can be delayed until the first expected downtime,
- situations where the object that can be kept in operation under special control.

In some cases, when, for example, there are no spare parts needed for the repair process or it is not possible to stop the operation, the repair is delayed. This decision, as well as the details related to the detected fault, are entered into the database. When fault repairs are completed, the work order record may indicate that individual failures, for whatever reason, could not be solved, which should also be entered into the database.

Repairs that have been completed are recommended to be re-inspected thermally. This retesting may also indicate that some faults have not been completely rectified, either due to improper installation, adjustment, poor construction, or factory failure. This result is re-entered into the database and the cycle can be repeated.

4 New Approaches in the Field of Power Transformers Maintenance

Recent studies in the field of transformer monitoring and diagnostics are generally focused on automated diagnostic systems, i.e. those which would be able to make a conclusion based on all available data. Those systems would then give credible advice on any possible problem.

An example of these monitoring systems based on dissolved gas analysis (DGA) is described in [15]. This system is uses the Duval Triangle method to determine the transformer state and fault type. Generally, stand-alone functions have limited success.

Neural networks (NNs) are also widely used in the field of transformer monitoring and fault analysis. Some of the commonly used methods include:

- Multi-Layer Perceptron NN [16],
- Back Propagation NN [17],
- Granular Computation NN [18],
- Cerebellar Model Articulation controller (CMAC) NN [19],
- Learning Vector Quantization NN [20].

Additionally, other intelligent systems such as fuzzy logic systems [21], decision trees [22], Support Vector Machine [23], k nearest neighbor (KNN) [23], evolutionary algorithms [24] and many more.

Generally, as stand-alone techniques show limited success due to their many restrictions, more combined techniques are used. These systems are known as hybrid systems

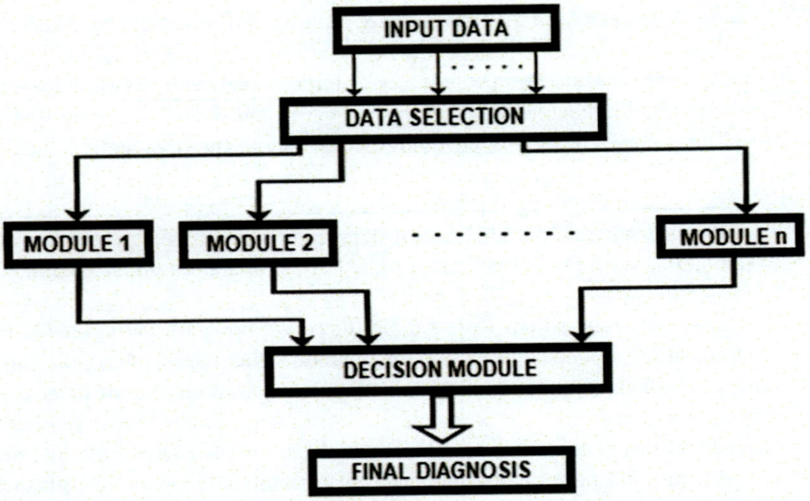

Fig. 6. Architecture of a hybrid diagnostic system.

where a different number of techniques are used to mitigate the negative effects of each individual technique. Figure 6 shows an example of a hybrid system architecture.

The general assessment of the diagnostic systems is based on its accuracy to determine state and/or fault correctly. It has been observed that systems with neural networks achieve better accuracy than other systems. This is due to the NN ability to learn nonlinear relationships between input and output data. Given the fact that the number of types of power transformers is somewhat large, it is reasonable to doubt the general applicability and reliability of a sole neural network only tested on only a relatively small data set.

On the other hand, DGA interpretation methods, which have been used in practice for over thirty years, have been developed based on the analysis of a large number of diagnosed transformer states with associated gas concentrations and have been applied to hundreds of thousands of examples worldwide. It is therefore certain that they cover the full range of types of power transformers [25]. However, all of these methods have relatively weak classification power so different methods and their combination can possibly show different results. These result can even be contradictory. Different methods using the same data set may conclude different fault types, recognize normal operating condition or even fail to conclude anything.

Tables 2 and 3 show some examples of the same data on gas concentrations using eight interpretation methods (IEC78, IEC99, MDT, RG3, RG4, KG, LN and DB).

Interpretation of DGA results is an extremely demanding procedure because, in addition to the application of an automated diagnosis method, it also requires extensive users experience. The DGA method achieves good results in the transformer condition analysis by combining interpretive methods as well as user knowledge and experience.

Due to the complexity of the fault mechanisms that are detected, as well as due to the complexity of applying the interpretive DGA methods themselves it is extremely

Table 2. Transformer fault types and distinct diagnosis over different methods [25].

Diagnosis	IEC78	IEC99	MDT	RG3	RG4	KG	LN	DB
NF	D1	ND	DT	D2	ND	ND	ND	NF
NF	ND	ND	T3	ND	ND	ND	ND	NF
PD	PD	T1	T1	PD	PD	PD	DT	NF
PD	ND	T1	PD	ND	D1	PD	T	ND
T1	T1	ND	T2	T1	T3	ND	DT	NF
T1	NF	T1	PD	NF	NF	PD	DT	ND
T2	T1	T1	T3	ND	T1	ND	T	T
T2	ND	ND	DT	ND	T3	ND	T	T
T3	T2	T2	T3	T2	ND	ND	T	T
T3	T3	T2	T2	T3	T3	ND	T	T
DT	ND	ND	D2	ND	ND	T3	D2	ND
DT	ND	ND	DT	ND	T3	ND	T	ND
D1	D1	D1	D2	ND	D2	PD	D2	ND
D1	ND	ND	T2	ND	ND	PD	ND	NF
D2	ND	T1	D2	ND	ND	D	D2	ND
D2	ND	T1	D1	ND	ND	PD	D2	NF

Table 3. Content of the diagnosis indexes [15].

Symbol	Diagnosis
NF	Normal state of the transformer
PD	Partial discharges
T	Thermal fault (regardless of temperature)
T1	Thermal fault $t < 300$ °C
T2	Thermal fault 300 °C $< t <$ 700 °C
T3	Thermal fault $t > 700$ °C
DT	Mixed thermal and electrical fault
D	Discharge (low or high energy)
D1	Low energy discharge
D2	High energy discharge
ND	The diagnosis cannot be determined

unlikely that there will be a fully automated method for transformer condition assessment and fault detection.

5 Conclusion

Transformer monitoring systems have found their application in power stations around the world and proved their importance in terms of preventing transformer faults, protecting staff and the environment, and better transformer management, which was especially evident after the liberalization of the electricity market. Knowledge of the state of the transformer, not only at the time of occasional diagnostic tests, but also during its operation, is becoming increasingly important. In order to make the best use of the monitoring system data, automatic diagnostic systems are being developed. On-line diagnostics of transformers are still mostly in the research phase and have yet to be confirmed in practical application. If its application is deemed promising, it is to be expected that its results and recommendations will not be a direct trigger for decision-making, but an auxiliary tool in making decisions about transformer management.

Thermal imaging tests are carried out preventively, with the aim of early detection of the causes of dangerous warming. In this way, it is possible to intervene in a timely manner and prevent forced downtimes, major damage, accidents, and in connection with this, large financial costs.

The best results can be achieved by applying automated diagnostics based on DGA results. However, the key problems to be solved in the development of a DGA-based automated diagnostic system are the development of a method for evaluating diagnostic findings obtained by interpretive methods and the development of a method for making a final diagnosis. The combination of the method for evaluating diagnostic findings and the method for making a final diagnosis must be effective enough to sufficiently compensate for the knowledge and experience of experts who are key elements of traditional application of DGA interpretation and are not available in automated diagnostics.

References

1. Sturm, F.A.: Efficient operations—intelligent diagnosis and maintenance. VGB Power Tech e.V., vol. 1003, pp. 195–303. Essen (2003)
2. Keith Mobley, R.: An Introduction to Predictive Maintenance; Butterworth Heinemann. New York (2002)
3. Banović, M., Keitoue, S.: Enhanced remote access to the monitoring system. In: International Conference on Condition Monitoring and Diagnosis CMD 2006, paper 659. Changwon, Korea (2006)
4. Banović, M., Keitoue, S.: Mobility support for the access to a monitoring system. In: IMEKO XVIII World Congress Brazil, Rio de Janeiro (2006)
5. Dropulić, T., Banović, M., Keitoue, S.: Inteligentno upravljanje rashladom transformatora sustavom motrenja transformatora Končar TMS. HO CIGRE 7. simpozij o sustavu vođenja EES-a, Cavtat (2006)
6. CIGRE Brochure 642: Transformer Reliability Survey, Final report of working group A2.37 (2015)
7. Nicola, M.: Monitoring system for power transformer windings hot spot temperature using fiber optic sensors, Kalman filter and integration in SCADA system. Am. J. Signal Process. **8**(2), 33–44 (2018)
8. McCalley, J.D., et al.: Automated integration of condition monitoring with an optimized maintenance scheduler for circuit breakers and power transformers. Final Project Report, Power Systems Engineering Research Center (2006)

9. Bengtsson, C.: Status and trends in transformer monitoring. IEEE Trans. Power Delivery **11**, 1379–1384 (1996)
10. Davidenko, I.V.: The development of diagnostics criteria of power transformers and instrument transformers using the results of the chromatograph analysis of gases dissolved in oil. In: International Conference on Condition Monitoring and Diagnosis CMD 2006, paper 839. Changwon, Korea (2006)
11. Akbari, A: Dissolved gas analysis based on information fusion to enhance reliability of power transformers diagnosis. In: International Conference on Condition Monitoring and Diagnosis CMD 2006, paper 597. Changwon, Korea (2006)
12. He, Q., Si, J., Tylavsky, D.J.: Prediction of top-oil temperature for transformers using neural networks. IEEE Trans. Power Delivery **15**, 1205–1211 (2000)
13. Sparling, B.D.: Moving forward from monitoring to diagnostics. IEEE/PES Trans. Distrib. Conf. Exposition **2**, 960–963 (2001)
14. Chandavong, P., Wiroteurairuang, D.: The development of power transformers diagnosis framework using infrared thermograms analysis. Int. J. Inf. Electron. Eng. **8**(1) (2018)
15. Duval, M., Dukarm, J.: Improving the reliability of transformer gas-in-oil diagnosis. IEEE Electr. Insul. Mag. **21**(4), 21–27 (2005)
16. Salami, A., Pahlevani, P.: Neural network approach for fault diagnosis of transformers. In: 2008 International Conference on Condition Monitoring and Diagnosis (2008)
17. Wang, Y., Zhang, L.: Transformer fault diagnosis based on back-propagation neural network optimized by cuckoo search algorithm. In: 2017 3rd IEEE International Conference on Control Science and Systems Engineering (ICCSSE) (2017)
18. Maumela, J., Nelwamondo, F., Marwala, T.: Condition monitoring of transformer bushings using rough sets, principal component analysis and granular computation as preprocessors. In: 2013 International Conference on System Science and Engineering (ICSSE) (2013)
19. Pai, N.-S., Lai, Y.-C.: Fault diagnosis of electric power system transformer on CMAC neural network approach. In: 2012 International Conference on Fuzzy Theory and Its Applications (iFUZZY2012) (2012)
20. Liu, J., Liang, Y., Sun, X.: Application of learning vector quantization network in fault diagnosis of power transformer. In: 2009 International Conference on Mechatronics and Automation (2009)
21. Wu, T., Tu, G., Bo, Z.Q., Klimek, A.: Fuzzy set theory and fault tree analysis based method suitable for fault diagnosis of power transformer. In: 2007 International Conference on Intelligent Systems Applications to Power Systems (2007)
22. Zhao, F., Su, H.: A decision tree approach for power transformer insulation fault diagnosis. In: 2008 7th World Congress on Intelligent Control and Automation (2008)
23. Benmahamed, Y., Teguar, M., Boubakeur, A.: Diagnosis of power transformer oil using PSO-SVM and KNN classifiers. In: 2018 International Conference on Electrical Sciences and Technologies in Maghreb (CISTEM) (2018)
24. Hong-xia, X., Li-ping, S., Zheng-yun, H., Hui, X.: Power transformer fault diagnosis based on chaos immune evolutionary clustering algorithm. In: 2010 2nd International Conference on Signal Processing Systems (2010)
25. Banović, M.: Automatska dijagnostika energetskihtransformatora, kvalifikacijski doktorski rad, FER

Requirements for High-Quality Thermal Inspection of the Transmission Lines

Igor Lukačević and Domagoj Lagator(✉)

HOPS d.d, Vukovarska Cesta 217, 31000 Osijek, Republic of Croatia
domagoj.lagator@hops.hr

Abstract. The paper deals with the basic characteristics of a thermographic camera important for the thermographic inspection of transmission lines. The specifics of the transmission line and the facts that can lead to incorrect measurement results are listed. Several thermographic images with a hand-held camera and a drone are shown. Each recording was analyzed, and possible shortcomings in the recording were listed. In conclusion, the minimum characteristics of hand-held thermographic cameras and thermographic cameras on drones are stated, as well as the recording conditions for correct thermographic inspection of transmission lines.

Keywords: Thermography · Transmission line · Thermal imaging camera · Drone

1 Introduction

In recent years, thermal imaging examinations have been an important method for diagnosing faults in high-voltage transmission lines. Transmission lines with nominal voltages of 110 kV, 220 kV and 400 kV are now regularly inspected with an IR camera every five years or as needed. The target inspection points are all junctions in transmission lines.

Thermal imaging inspections of transmission lines began to be regularly monitored ten years ago. Previously, these inspections were not performed as regular work because they were not prescribed as mandatory inspections in the internal rules of the electricity company. Such examinations were performed only in exceptional situations based on various indications. Due to increasing requirements for the availability of transmission lines, increasingly strict conditions for disconnecting certain sections of transmission lines and requirements to reduce maintenance costs, the deadlines for regular periodic maintenance (revision of transmission lines) have been extended. In order to maintain, or even increase, the reliability of the drive, regular periodic inspection of transmission lines with an IR camera was put into practice. The intended frequency of inspection is every five years. Inspections refer only to connection points on the transmission line, which are current bridges on tension poles, or repair connectors in places where the conductor has been changed or repaired.

Thermal imaging inspection of transmission lines has several specific features compared to inspections of other power plants:

D. Blažević et al. (Eds.): OTO 2022, LNNS 592, pp. 84–95, 2023.
https://doi.org/10.1007/978-3-031-21429-5_8

- the recording object (coupler) is relatively small and distant,
- the current load of transmission lines is difficult to influence,
- field nature of the work.

Due to the limited resolution of the camera sensor, the size of the object being recorded affects the accuracy and reliability of the results.

To avoid this problem, the cameraman must

- get as close as possible to the object,
- use a higher resolution camera,
- or use a lens with a narrower angle (telephoto lens).

To increase the possibility of spotting hot spots, the load on the transmission line should be as high as possible, which is difficult to influence on the transmission lines. Also, due to the development of modern technologies, inspections of power lines from the air have recently been carried out by unmanned aerial vehicles.

2 Technical Characteristic of IR Camera

2.1 Resolution

The resolution of digital cameras is one of their most important characteristics. The resolution of the camera sensor defines the number of pixels that convert the image into an electrical signal or a digital image. In the visible part of the spectrum we are talking about video resolution, and in the infrared part of the spectrum we are talking about thermal resolution. Given today's technological development in the field of digital cameras and video cameras, resolutions from ten to even one hundred megapixels are understood. In the field of IR devices, these values are much smaller. High-quality commercially available devices for civilian use generally have a resolution of up to 0.3 megapixels.

The resolution is most often expressed according to the size of the matrix in the number of pixels in the x and y directions. Today's IR cameras usually have a resolution of 160 × 120, 320 × 240, 640 × 480 and 640 × 512 pixels.

2.2 Lens

The role of the thermal camera lens is the same as the role of the lens of an ordinary camera, i.e. it focuses thermal radiation on the matrix of thermal sensors. However, thermal imaging camera lenses are made of germanium because ordinary glass is not permeable to infrared radiation. For this reason, the price of thermal lenses is several times higher than that of visible light lenses (Fig. 1).

The basic characteristic of the lens is the angle of view (AOV) and field of view (FOV). The usual viewing angle of a thermal camera is 24°. However, more complex and better quality cameras allow changing lenses. Lenses ranging from 7° (telephoto lenses) to 45° (wide-angle lenses) can thus be used.

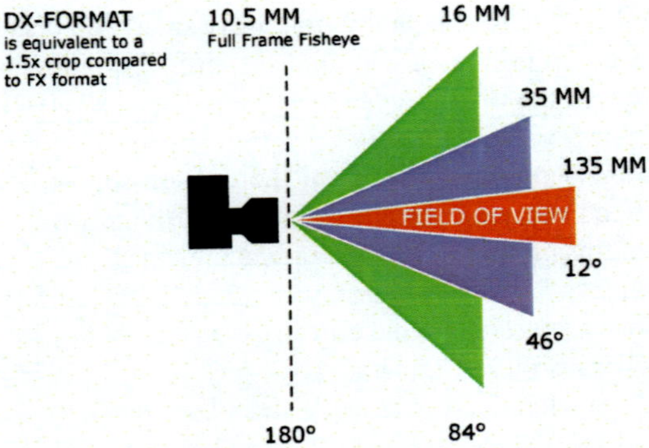

Fig. 1. Illustration of camera resolution and field of view

From the resolution of the camera and the field of view of the lens comes the most important characteristic of the camera, which is the field of view of one pixel (IFOV—Instantaneous field of view). Let's assume that the thermal camera has a resolution of 640 × 480 pixels and has a lens with an angle of 24°.

The IFOV parameter is calculated [1]:

$$\begin{aligned} 24^\circ/640\ \text{pix} &= 0.0375^\circ \\ &= 0.00065\ \ \text{rad} \\ &= 0.65\ \ \text{mrad} \end{aligned}$$

So one pixel covers an angle of 0.65 mrad, which means that at a distance of one meter it covers an area of 0.65 mm.

For ideal camera optics, it would be sufficient if the observed object was larger than the area covered by one pixel for the temperature reading to be accurate (Fig. 2a). However, it is not certain that the pixel will always be centered on the object (Fig. 2b). In this case, the measured temperature will be the average temperature of the object and the ambient temperature. Measurement error is certainly present if the observed object is smaller than the pixel size (Fig. 2c) [1].

Given that the optics of the lens are not perfect, part of the thermal radiation is scattered by passing through the lens. In this way, a diffused image is created around the pixel and all the radiation does not fall on it. Therefore, a larger number of pixels per object is required to avoid measurement error. The area covered by a pixel for correct measurement is denoted by MFOV (Measurement Field of View or True IFOV) (Fig. 3) [1].

For modern cameras with good optics and taking into account the above, the size of the observed object should not be smaller than the area covered by 3 × 3 pixels.

2.3 Other Characteristics

In addition to the above-mentioned characteristics, each camera has a whole series of characteristics related to recording itself, ergonomics, functionality, specific purpose, robustness, software characteristics, etc. We will single out just a few:

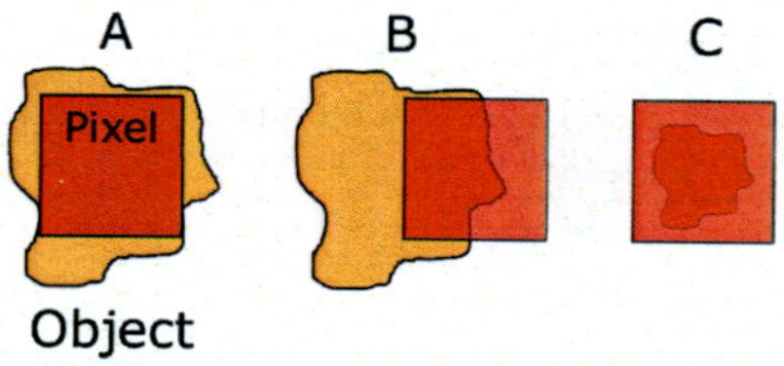

Fig. 2. Illustration of the relationship between pixels and the observed object [1]

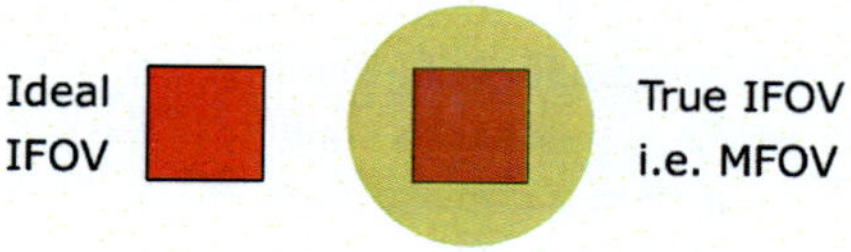

Fig. 3. Scattering of radiation around an ideal pixel [1]

Thermal sensitivity or Noise Equivalent Temperature Difference (NETD) describes the smallest temperature difference you can see with the camera. Typical values are 100 mK for worse models to 30 mK for better models. Figure 4 shows the difference in thermal images for two different thermal sensitivities.

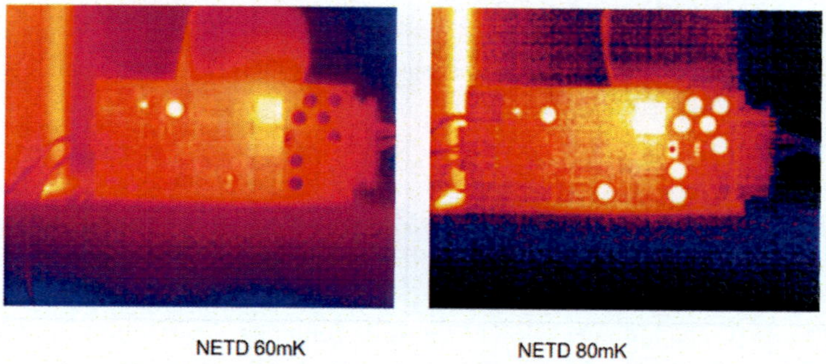

Fig. 4. Thermal images made by cameras with different thermal sensitivities

Zoom—IR cameras, as a rule, do not have the possibility of optical zoom, but the image enlargement is performed by software. This way of enlarging the image cannot improve and increase the accuracy of the measurement, but it can help in noticing details and more precise focus/sharpening of the image [2].

Focus—In lower class cameras, the focus or sharpening of the image is achieved automatically. With higher class cameras, focus can be done automatically or manually. When using a telephoto lens (with a smaller viewing angle) and recording smaller and more distant objects, manual focus adjustment is necessary [2].

Refresh rate—The refresh rate of the thermal image is important for recording moving objects and when you want to make a thermal video. Refresh rates range from 7 Hz to 30 Hz [2].

Software image enhancement—By applying various methods of interpolation and digital image processing, image resolution can be increased (Fig. 5). The correctness of the application of these methods depends on the assumed physical model. Since heat propagation is a spatially continuous phenomenon without sudden discontinuities, the surface interpolation of the measured values makes sense (as in the example in Fig. 5). However, when imaging a spatially isolated object such as an electrical conductor or coupling in air, there is a large discontinuity in the temperature of the observed object and the background. Namely, in this case the background is the sky with an extremely low radiation temperature, which in clear weather is around −70 °C (−46 °C in Fig. 6). In this case, applying any interpolation can lead to incorrect temperature readings [3].

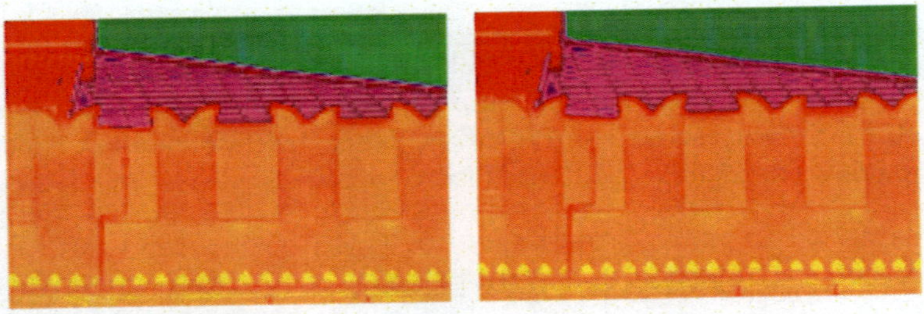

Fig. 5. Software enhancement of thermal image resolution

3 Specifications of Thermal Vision Inspection of Transmission Lines

Thermal imaging inspection of transmission lines is performed in order to identify bad connection points where excessive heating of conductors and connectors occurs. Therefore, the connectors of the phase conductors are primarily inspected. These couplings can be of compression or screw type and are located primarily on tension columns. In addition to the tension poles, teaching and repair couplings can be located anywhere in the transmission line route.

Let's list some basic peculiarities of recording these objects:

- The size of the observed object is relatively small considering the distance. The diameters of the phase conductors of transmission lines range from 16 to 31 mm.
- The distance of the measuring device (camera) from the object is relatively large. For practical reasons (high altitude, accessibility of the route) and safety (high voltages), the measurement is made from a distance that is rarely less than 20–30 m. In recent years, when recording from drones, this value is much smaller. The average distance when recording from a drone is about 3–10 m.

- The background of the object is extremely low temperature (atmospheric radiation temperature, depending on the weather conditions, drops to—70 °C).
- Manipulations with the switching state of the network can hardly increase the load on the transmission line in order to facilitate the detection of a hot spot. As a rule, power lines are under low load at the time of recording.

According to Sect. 2.2. Let's calculate the maximum angle of the camera lens.

- assume the value of MFOV is 3 pixels and the thickness of the conductor is 20 mm
- it follows that 1 pixel must cover an area of 20 mm/3 = 6667 mm

At a distance of 30 m from the object, it follows the:

- IFOV = 0.006667 m/30 m = 0.00022 mrad.

For a camera with a resolution of 640 × 480 pixels, we calculate the required viewing angle of the lens

- 0.00022 mrad * 640 pix * 180°/π = 8149°. (standard lenses on the market have an angle of 7°).

Figure 6 shows a correct thermal image of tension clamps on a 110 kV transmission line made with a camera with a resolution of 640 × 480 pix and a 7° lens from a distance of approx. 25 m. When using a 7° lens, attention must be paid to sharpening the image. The maximum sharpness of the image is achieved by manually adjusting the focus of the lens with the mandatory use of a geodesic or photographic tripod. Figure 7 shows a defective thermal imaging image due to poor image focus, regardless of the fact that the correct equipment was used (camera 640 × 480 pix, lens 7°).

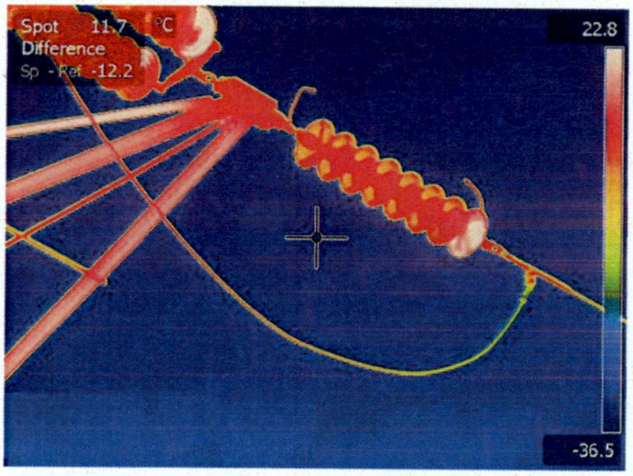

Fig. 6. Thermal image of tension clamps on a 110 kV transmission line made with a camera with a resolution of 640 × 480 pix and a 7° lens from a distance of 25 m

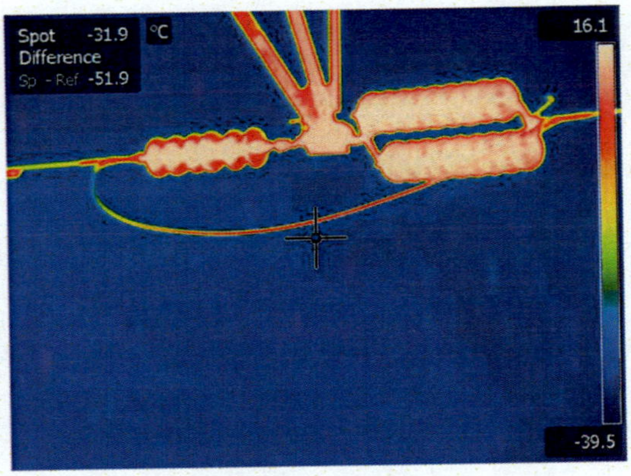

Fig. 7. Defective shot due to poor image focus

Due to the development of technology recently, unmanned aerial vehicles are also used for inspections of power lines, including thermal imaging inspections. The advantage of this kind of inspection is that the object can be approached at a shorter distance, regardless of the accessibility of the route. The disadvantage is that the payload of drones is limited and large cameras cannot be used. Regardless, technology has managed to assemble cameras of compact dimensions for unmanned aerial vehicles that can provide the same quality image as much more expensive cameras for recording from the ground.

An example is a camera with a resolution of 640 × 512 pix and a wide-angle lens of 40.6°. Let's calculate to what distance from the 20 mm diameter guide the unmanned aerial vehicle with such a camera must approach for a correct thermal image.

$$\text{IFOV} = 40.6^{\circ}/640\ \text{pix} * \pi/180 = 0.00111\ \ \text{mrad}$$

With the assumed MFOV of 3 and a conductor thickness of 20 mm, the maximum distance d is

$$\text{d} = 0.020\ \ \text{m}/(0.00111\ \text{mrad} * 3) = 6.01\ \ \text{m}$$

Approaching such small distances requires great attention and experience of the pilot, as well as the resistance of the unmanned aerial vehicle's equipment to electromagnetic influences. Also, it is difficult for the pilot to judge the actual distance of the aircraft from the object.

Figure 8 shows a thermal image of a 110 kV transmission line made by an unmanned aerial vehicle with a 640 × 512 pix camera and a 40.6° lens. The picture was taken from a distance of about 3 m. The advantage of shooting with an unmanned aerial vehicle is an easy check whether it is really a warm place or a reflection. Shooting with a camera requires a lot of moving and checking that takes a long time. Shooting with an unmanned aerial vehicle enables a quick change of the shooting angle and repeating the shot of the object.

Subsequently, a power line team was sent to the field to determine if it was really a warm place. The transmission line was disconnected and the recorded current bridge was dismantled. After disassembly, it was evident that the recording was correct and that the warm spot was correctly detected.

Fig. 8. Thermal image of a 110 kV transmission line made by an unmanned aerial vehicle with a 640 × 512 camera and a 40.6° lens

The next few pictures show thermal imaging images made in an incorrect way and perhaps with wrong conclusions. Figure 9 shows a thermal image of a 110 kV transmission line in the coastal area. The image is of low resolution and it is difficult to draw a correct conclusion from it. The temperature reading of the connectors shows a temperature difference of about 20 °C. Given that the recording was made in March in the coastal area, it is not realistic that the conductor temperatures are below 0 °C, especially if it is compared to the temperature of the structure. The low conductor temperature reading is the result of averaging the actual temperature with the extremely low sky temperature.

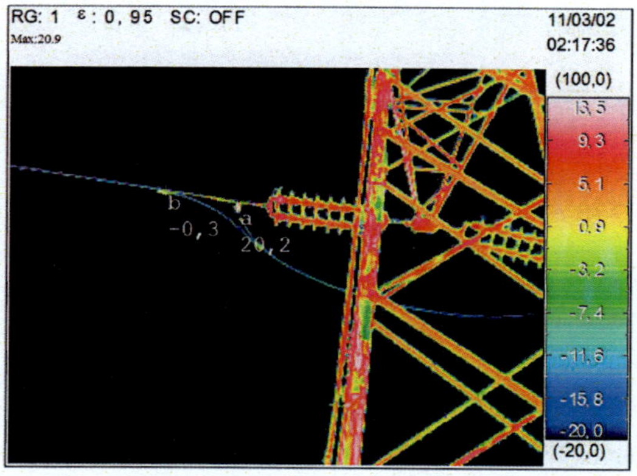

Fig. 9. Thermal image of 110 kV transmission line in the coastal area made with a 320 × 240 pix. camera

Figure 10 shows a thermal image of a column where potentially bad connection points can be seen. Due to the chosen small temperature range of the image (0.3 °C to 6.3 °C), the observed points appear extremely warm, but the real difference compared to the ambient temperature is approx. 4 °C. It is also unlikely that all four clamps shown are faulty. Namely, there is no electrical connection at all on the clamps shown, i.e. conductor breaks.

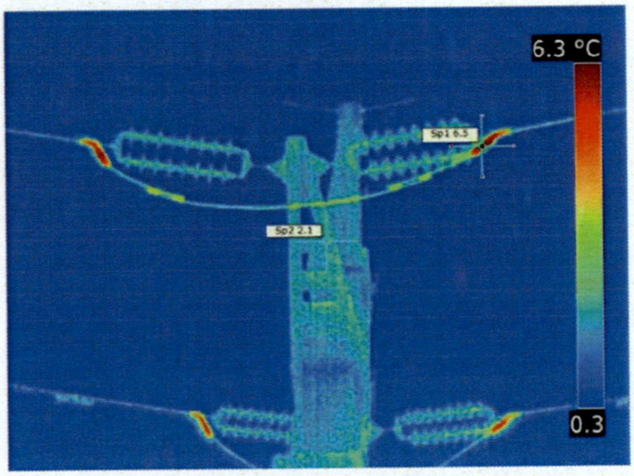

Fig. 10. Thermal imaging image of a 110 kV transmission line pole with insufficient resolution

Figure 11 shows a shot of a 400 kV transmission line with a good resolution camera (640 × 480 pix) but with a wide angle lens (24°) and from a long distance. Nothing can be concluded from this picture about the condition of the connection points on the transmission line.

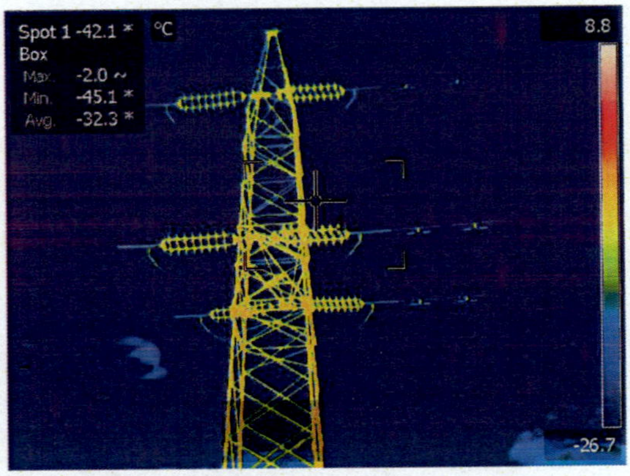

Fig. 11. Thermal image of a 400 kV transmission line with a camera of good resolution (640 × 480 pix), but with a lens with a wide angle (24°) and from a long distance

4 Comparison of the Specifications of a Handheld Camera and a Drone Camera

Recently, there has been a lot of discussion about which camera is better to use in work and which gives better measurement results. Like everything, this depends on the situation, but in general it can be said that both cameras give excellent results when used within their operating parameters and when all the rules of thermal imaging are followed.

In transmission system operator, we have been using the IR camera FLIR P660 in combination with a narrow-angle 7° lens for thermal imaging examinations for a long time. Two years ago, we also started using an IR camera carried by an unmanned aerial vehicle. It is a DJI H20T camera.

A hand-held camera has its advantages when recording power plants, because drone flights are not possible in these areas. Likewise, these are extremely accessible spaces and it is easy to move the camera and take quality shots.

Hand-held cameras also have their advantages when recording isolated transmission line poles that are extremely accessible, that is, those poles that can be reached by vehicle. The reason for this is that the preparation of the drones, as well as the preparation after the flight, still requires a certain amount of time.

If the pole is on inaccessible terrain, or if it is in the middle of some culture, that is, if the pole is in a fenced area, or if we have several tension poles in a row, then the drone camera wins.

For example, 6 poles of a 400 kV two-system transmission line were recorded from one takeoff with an unmanned aerial vehicle (a total of 36 points for checking). This would require a minimum of 12 moves using a handheld camera to capture everything and it would be necessary to reach the base of all 6 towers (Table 1).

Using the calculation from Sect. 3, in Table 2 we will give the maximum distances from which it is possible to obtain a high-quality recording when recording conductors and equipment on portable transmission lines.

When filming with a hand-held camera, we can say that as a rule we always shoot from a limited distance because it is often difficult to approach each column to its base.

When filming with a hand-held camera, we can say that as a rule we always shoot from the limit distance because it is often difficult to approach each pillar to its base. When filming with an unmanned aerial vehicle, we are always below the limit distance, it is rarely filmed from a distance that is close to the limit. Another advantage of an unmanned aerial vehicle is that it is possible to see the exact distance of the object from the aerial vehicle at any time, while this is not possible when recording with a camera unless some distance measurement equipment is additionally worn.

5 Conclusion

Based on the above, it can be concluded that for a reliable thermal imaging inspection of transmission lines by recording from the ground, a camera with a resolution of at least 640 × 480 pixels and a lens with an angle of 7° should be used, provided that the location of the recording is as close as possible to the object. Likewise, aerial photography is also

Table 1. Comparison of the two types of cameras used in the event thermal imaging of transmission lines in the transmission system

	Flir P660 + 7° lens	DJI H20T
Resolution	640 × 480 pix	640 × 512 pix
Lens	24° (serial), 7° (optional)	40.6°
Temperature sensitivity	0.045 mK	0.050 mK
Temperature range	−40 °C do 500 °C	−40 °C do 550 °C
Zoom	digital 8x	digital 8x
Refresh rate	30 Hz	30 Hz
RGB camera	built-in 3.2 Mpix camera does not work when using the 7° lens	built-in camera 20 Mpix, 23x optical zoom
GPS	yes	yes
Post-processing of the image	yes	yes
Maximum distance when recording conductors with a diameter of 20 mm	35 m	6 m
IP protection	None	IP44

Table 2. Comparison of the maximum distances for creating a quality thermal image.

Conductor type (ACSR) (mm^2)	Flir P660 + 7° lens (m)	DJI H20T (m)
490/65	53	9
360/57	46	8
240/40	38	6.5
150/25	30	5

possible. All cameras with a resolution higher than 640 × 480 are good, as well as any camera with a shooting angle of less than 40.6°.

In addition to these specifics of transmission line imaging, other conditions for high-quality thermal imaging must also be considered. The basic requirement is a well-trained videographer with experience in interpreting thermal imaging images. It is also necessary to know the technical characteristics of the recorded object, types of materials and emission factors, etc., but this is not the subject of this paper.

References

1. Gustavsson, R.: Thermography: A Practical Approach. NORBO KraftTeknik AP (2009)
2. Flir homepage. https://www.flir.com/discover/professional-tools/thermal-camera-specs-you-should-know-before-buying/

3. Mandanici, E., Tavasci, L.: A multi-image super-resolution algorithm applied to thermal imagery (2019)
4. Jaman, N., et al.: Croatian transmission system operator Messenger 5, Zagreb (2016)
5. Bojan Milovanović, Flir ITC (Infrared Training Center) Level 1, Faculty of civil engineering Zagreb, November 2018
6. Movitherm homepage: https://movitherm.com/knowledgebase/netd-thermal-camera/
7. Shotkit homepage. https://shotkit.com/field-of-view/

The Influence of Preventive Maintenance of Batteries on Increasing the Security of the Thermal Power System

Duška Jarčević[1], Ljiljana Radovanović[2](✉), Jasmina Pekez[2], Borivoj Novaković[2], and Hrvoje Glavaš[3]

[1] Panonske Termo-Elektrane—Toplane Zrenjanin, Zrenjanin, Serbia
[2] Technical Faculty "Mihajlo Pupin", University of Novi Sad, Zrenjanin, Serbia
ljiljana.radovanovic@tfzr.rs
[3] Faculty of Electrical Engineering, Computer Science and Information Technology Osijek, University of Osijek, Osijek, Croatia

Abstract. Rechargeable batteries are historically the oldest chemical source of electricity, and are still used today as a backup source of electricity to support power plants operation. The batteries connected to the auxiliary DC voltage busbars are powered from the AC mains by means of industrial electronics devices. The paper presents a study of the choice of appropriate battery maintenance methodology in order to increase the safety of thermal power systems, reduce system failure and increase reliability. A proposal for preventive maintenance through regular inspections and periodic replacement of individual cells is presented. The search for weak cells and their elimination is accompanied by technical diagnoses and tests of correctness and capacity. Finally, the results of the tests carried out at the Zrenjanin thermal power plant (the part of the company Panonske Termo-elektrane—toplane Zrenjanin) at five-year intervals are presented. In the concluding remarks, the need for maintenance is emphasized with a view to improving the system and reliability.

Keywords: Preventive maintenance · Batteries · Thermal power system

1 Introduction

While the AC mains is present, the batteries are in a state of constant recharging, and the DC consumers are powered by rectifiers. Sensitive consumers are most often associated with problems with the quality of electricity supply. They do not tolerate long periods without power. Even a short-term outage can lead to catastrophic consequences (failure of the operating room in hospitals, loss of data in computer centers, banks, failure of the control system at power plants, etc.). The simplest configuration of uninterruptible power supplies consists of a rectifier, backup power source (battery), static switch and inverter. The basic principle of uninterruptible power supply operation is to charge the backup power source (battery) via the mains rectifier. The battery then powers the inverter which converts its direct energy (DC) to alternating energy (AC) and as such delivers it to consumers [1].

D. Blažević et al. (Eds.): OTO 2022, LNNS 592, pp. 96–105, 2023.
https://doi.org/10.1007/978-3-031-21429-5_9

When the AC power goes out, rechargeable batteries take over the consumer's power, ensuring uninterrupted voltage. Rechargeable batteries that are permanently installed and maintained in a power plant are called stationary batteries. A sudden loss of power can significantly disrupt business, but this is not the only failure in the multitude of failures in the power grid that can have devastating consequences. The effects of inadequate power supply are endangering human life, endangering control systems in industry, business interruption, loss of software and data, destruction of hardware and its components, downtime, loss of telecommunications [2].

2 Basic Characteristics of Stationary Batteries

Stationary batteries are a source of direct energy for stationary devices (see Fig. 1). They are used for power supply of signal and command devices at various power supply stations of command devices, radio telegraph and telephone exchanges and others [3]. Stationary batteries also serve as backup power in the event of a power outage for emergency lighting and more, and are intended for use in fixed locations where it is not common to move from place to place and are permanently connected to be charged to a DC power source [2]. The battery plant is shown in Fig. 2.

Fig. 1. Lead acid rechargeable battery. *Source* TE-TO Zr, personal archive

The parts of a stationary battery are: positive electrode, negative electrode, separator, poles, box and electrolyte—dilute sulfuric acid. The stationary lead-acid battery consists of several cells connected in series and is used as a direct current power source.

Cell voltage—the nominal voltage of lead acid cells is 2 V. The voltage of a loaded cell of electrolyte density also depends on the load current, and as the current load increases, the cell voltage decreases. The ratio of cell voltage to load current depending on the size of the plates, during discharge with constant currents I10, I5, I3, I1.

Fig. 2. Battery plant. *Source* TE-TO Zr, personal archive

The capacity of the cell is expressed in ampere hours Ah and represents the amount of electricity given by the battery during discharge with a certain current for a certain time.

The service life of stationary cells with tubular plates is 1500 charge-discharge cycles or more than 10 years for stationary conditions of use.

3 Research Method

The main research problem in this paper relates to the need to increase the security of thermal power systems that rely in their work on providing direct and uninterruptible current for their proper functioning. Since rechargeable batteries serve as a backup source of electricity in power plants, their correctness is very important. Even a short-term outage can lead to catastrophic consequences. For these reasons, special attention should be paid to the maintenance of stationary batteries.

The subject of research in this paper relates to the application of preventive maintenance of batteries in order to increase their service life, improve the reliability of equipment and safety of stationary batteries on which all other systems rely.

The aim of the research is to define the best methodology for maintaining batteries, in order to improve the reliability of batteries, the entire plant where they are located.

Also, one of the goals of the research is to eliminate the problem of the occurrence of a large number of failures in the battery plant, and thus the entire thermal power system as a whole.

Namely, by determining the appropriate methodology for the maintenance of rechargeable batteries, multiple negative consequences would be eliminated in the event of failure of the entire plant, which is also the goal of the research.

The task of this research includes determining:

- The best methods of maintenance of rechargeable batteries in order to increase the safety of the system,
- Reducing the number of failures on batteries,
- Improving the reliability of rechargeable batteries by applying the chosen maintenance methodology.

Within this research, the methods of theoretical analysis was used (professional literature, project documentation and reports on battery measurements for the period 2011–2016 had been analyzed). To control the correctness of the battery plant, an experimental method was used using a testing technique that measures physical quantities (voltage, temperature, etc.). Then a comparative analysis was used to compare the values of physical quantities by years from 2011 to 2016.

The instruments of this research are the means used within the experimental research method: test, documents from the archives of TE-TO Zrenjanin and record sheets.

The following devices and instruments required for AB maintenance were used for testing stationary batteries: DC instrument measuring instrument; Thermometer at 100 °C; Accuracy class ammeter and with a set of shunts; Voltmeter of accuracy class at least 1.5 with measuring range 0–3 V with a division of 0.02 V; A clock; Density hydrometer with measuring range from 1.1 to 1.3 g/cm^3 [2].

The research was performed in the part of the company Panonske Termoelektrane—toplane Zrenjanin in the plant—the battery located in the hall in the annex of the main power plant (Fig. 2), which was designed in relation to the main power plant [2].

4 Proposal for Implementation of Preventive Maintenance Procedures Rechargeable Battery

4.1 Preventive Periodic Examinations

In addition to the ten-hour current test, the capacity test can be performed with a five-hour, three-hour and one-hour discharge current. Discharge currents as well as capacity depend on the construction of the battery and vary according to type and manufacturer. When determining the capacity, we are obliged to use the data provided by the manufacturer on the discharge currents and capacities for each of the test modes. Discharge itself is a loss of battery capacity due to internal chemical and electrochemical reactions when the circuit is open. Self-discharge of stored batteries is about 1% of the nominal capacity per day at an ambient temperature of 20 °C. This percentage decreases over time. Newly formed cells may be stored for up to one year [2].

On a daily basis, you should visit the battery room every day to visually determine if there are any defects. If the electrolyte level is below the level, fill the cell with distilled water, and in case of any other deficiencies, inform the competent management of the electrical maintenance service.

On a monthly basis, check the voltage on each cell once, if the voltage per cell is below 2.1 V, you should start recharging the battery. Also, clean the battery covers and the joints of the battery covers with the battery container with a dry cloth once a month. Clean and coat any oxidized contacts, check the room ventilation function.

Quarterly required: measurement of electrolyte density; electrolyte temperature measurement; electrolyte level control; visual inspection of plate and sediment accumulator vessels; cell voltage check; contact overview; visual inspection of the battery stand; checking the ventilation of the room; review labels and warnings.

Per year It is necessary to rinse the plugs annually and determine the capacity of all batteries on the control ten-hour discharge.

Regular maintenance should include: checking the cell voltage; electrolyte density check; partial discharge; battery overview [2].

The cell voltage should be checked once a month. Checking the cell voltage at the transformer station with two batteries—should be done so that the battery is disconnected from the consumer and the rectifier. The nominal voltage of the cell is 2 V (2–2.1). The average voltage of the cells will depend on the load current. In this case, it is necessary to reduce the consumer current (relieve consumption) as much as possible. Voltmeters used for measurement should have an internal resistance of at least 1000 Ω/V, measuring range 0–3 V jss and 0–250 V jss, accuracy class 0.5.

After the measurement of the voltage across the cells is completed, it is necessary to measure the voltage at the ends in order to check the connections and connection points [2].

Electrolyte density should be checked once a month. Rechargeable batteries use a solution of sulfuric acid with a density of 1.2400 ± 0.005 g/cm^3 at a temperature of 20 °C at nominal voltage. Electrolyte density decreases with increasing temperature and vice versa, with decreasing temperature, electrolyte density increases. If the density of one cell deviates by more than ± 0.05 g/from the average density of other cells, then it is defective, so it is necessary to check the density of electrolytes by voltage control and determine the malfunction of the cell. Measurement of electrolyte density (specific mass) is performed by AREOMETRIMS, measuring range from 1.00 to 1.300 g/with a division of at least 0.005 g/and an accuracy of at least 0.005 g. Since the density of the electrolyte depends on the temperature, it is necessary to measure the temperature of individual cells. This measurement is performed with a thermometer measuring range 0–100 °C and 0.5 °C. If the temperature of a cell deviates by more than 1 °C from the average temperature of other cells, it indicates its malfunction (along with other parameters) [2]. Partial discharge of the battery is performed once a month. This discharge is performed in order to prevent sulfation of cells "breaking" of solid sulfates deposited aku-battery cells. Discharge is performed by turning off the rectifier and leaving the battery to power consumers. Partial (preventive discharging) is performed with currents corresponding to 1–1.5 ten-hour discharge currents (I10–1.5 I10) because only these values of current provide decomposition melting of lead sulfate crystals. In case the current of connected consumers is below the value of I10, it is necessary to turn on additional consumers. Discharging lasts until the battery loses about 20–25% of its rated capacity. This partial discharge is preventive in nature and significantly prolongs the service life and reliability of batteries treated in this way [2].

By inspection, rechargeable battery is understood as a series of tasks in order to determine the general condition of the battery, the space in which it is located in the anteroom, ventilation system, etc. All tasks on the inspection of the battery and the deadlines for their performance are listed in Table 1. Checking the electrolyte levels and topping up the distilled water is done once a month [2].

Periodic maintenance activities include actions on batteries performed by teams trained for these tasks and performed according to the established plan. These tasks include: tightening the connections, deep emptying to check the capacity, forced charging. Deep emptying is done once a year. In the case of transformer stations with a single battery, a spare source of direct current power supply (spare mobile battery) must be provided during deep discharge. Discharge should be performed within one time, and at most 18 h after the end of the additional charge. During the discharge, the ventilation filter plugs on the cells must be removed. Measurement of electrolyte density and voltage of each cell is performed every hour. Charging of batteries must begin no later than 12 h after discharge. When the battery is empty, the AC adapter turns on the forced (deep) charge position. During this charging mode, the voltage is 2.6–2.7 V/cell. Then consumers must be excluded. Charging is completed when the final charging voltage specified by the manufacturer is reached, when the voltage has not changed during the last two hours, and the prescribed electrolyte density has been reached. At the end of the filling, the electrolyte levels were adjusted by adding distilled water. Ventilation filter plugs are flushed and replaced. The rectifier is left to run in refill mode or automatic mode [2].

4.2 Preventive Periodic Replacements

Criteria for replacement of individual cells are: cell voltage, cell electrolyte density and cell capacity. Procedures for checking cell voltage and electrolyte density, as well as the tolerance of the availability of measured values from the declared ones, are given in regular maintenance. One type of measurement alone is not enough to declare a cell defective. The capacity of the correct cell must be at least 70–80% of the declared value at a given discharge current. Criteria for the replacement of entire batteries, in addition to the technical ones that apply to individual cells (electrolyte density, voltage, capacity), contain other factors such as: the time the battery spent in operation and economic and commercial reasons. Manufacturers give the battery life as a function of the number of discharge cycles, which translated into service life is 14–15 years. It should be borne in mind here that the service life of batteries depends on the operating conditions and the quality of maintenance. The average service life of batteries in our plants is 10–12 years [2].

4.3 Finding and Fixing Vulnerabilities

If the discharge and charge voltage is reduced and the amount of electrolyte is reduced at the same time, it is an increased degree of sulfation or a short circuit—it is necessary to replace the cell.

If the cell temperature is increased, it is a process of sulfation in an advanced phase—it is necessary to conduct a training cycle which is carried out by discharging the AB with ten-hour current, until it loses 50% of capacity. It is then charged at a voltage of 2.4 V/cell. If such cycles are reduced to the correct state, if the discharge voltage and capacity during deep discharge are reduced, it is a process of intensive sulfation and it is necessary to conduct a training cycle. It is an electrolyte contaminated with foreign impurities—electrolyte replacement is required. If the electrolyte leaks, the box

Table 1. Battery inspection tasks and deadlines

Serial number	Description of works	Maintenance deadlines	
		Regularly	Periodically
1	Checking the battery voltage (basic and additional branches) and consumer current	Weekly	
2	Checking the cleanliness of the battery compartment and anteroom	Weekly	
3	Checking the electrolyte density of each cell	Monthly	
4	Checking the electrolyte levels of all cells and topping up	Monthly	
5	Check the voltage of all cells	Monthly	
6	Partial discharge of the battery, depriving it of 20–25% of the nominal capacity	Monthly	
7	Checking the condition of slabs and sediments	Monthly	
8	Overview of connections, compounds and cabinets with battery fuses	Monthly	
9	Checking HTZ equipment accessories and water distillation reserves		Every three months
10	Cleaning and washing the external surfaces of the battery		Yearly
11	Cleaning and renewal of Vaseline coatings		Yearly
12	Checking and cleaning ventilation openings and drains		Yearly
13	Tightening the battery connections		Yearly
14	Deep discharge of the battery to check the capacity		Yearly
15	Forced deep filling		G

Source [2]

is damaged. The battery-cell needs to be replaced. The most common errors that occur during battery operation are: sulfation; short circuits in the cell; electrolyte impurities; short circuits between cells; depolarization [2].

4.4 Technical Diagnostics—Testing the Correctness and Capacity of the Battery

Testing the correctness and capacity of batteries is shown in the report of the competent organizations by measuring ten hours of discharge current with constant monitoring of voltage and electrolyte in each cell of the battery during discharge, one correct lead stationary battery of 220 V. The testing is performed by the company Melbat Novi Sad.

The 80 OpzS 800 battery has 800 Ah 220 V and the test is performed with a discharger with a ten-hour current I10 of 80 A.

Before starting the measurement, the voltage of each cell and the density of the electrolyte in each cell are measured separately (first and second columns in the report) where the initial voltage of each cell and the total cell voltage of the whole battery are seen. The operating temperature of the medium is measured at 13 °C in the room of the plant with batteries. After that, the battery is discharged and after the first ten minutes, the first control measurement of the voltage of each cell is performed, and then every hour from the beginning of the test, the voltage of each cell and the total voltage of the whole battery are measured. In the following hours, a voltage drop is observed for each cell and when the voltage of each cell falls below 1.83 V, the measurement is interrupted because it is predicted that in the following hours the voltage will fall below 1.80 V per cell. After testing this battery—8 OpzS 800 A 220 V, it was determined that all ten hours the voltage per cell was above 1.80 V and has a capacity value of 100%. At the end of the tenth hour, the electrolyte density of each cell is measured [2]. Visual inspection checks the condition of the cells, as follows: there should be no lead sulfate sediment in the sedimentation tanks at the bottom of the vessel; the color of the correct positive plates (active mass PbOz) should be dark brown, and the color of the correct negative plates (active mass of sponge lead) should be light gray; correct plates must be flat and without deformations; separators made of special plastic foils and located on the surface of the plates, must rest on the plates; the spacers between the plates must be in place; the connections between the "collecting comb" plates as well as the floor leads must be free of cracks and damage [2]. The handling of batteries or products with an integrated battery must in all circumstances be in accordance with the manufacturer's instructions. This applies in particular to the limitation of the thermal load during storage and transport [4].

In the case of batteries with transparent containers, visual inspection is performed directly, in the case of semi-transparent containers by lighting with a flashlight through the container, and in the case of opaque containers through the ventilation plug opening. Cell surfaces must be wiped because dust or distilled water spilled during refilling significantly increases the self-discharge current [2].

Store rechargeable batteries or products with integrated rechargeable batteries, preferably at room temperature and in a dry place; avoid large temperature differences [4].

4.5 Checks

When bringing the battery into operation and charging, the room must be naturally ventilated or ventilation must be switched on. Access to open flames in the room is prohibited—batteries, smoking or the use of devices and tools that can cause sparks. Workers must wear special clothing, goggles, rubber aprons, gloves and boots when replacing individual cells, topping up and other work with electrolyte [2].

5 Research Results

In solving the problems of research and implementation of the set goals, the methodology of preventive maintenance of batteries was chosen, which includes the following procedures:

- Basic maintenance by the operator (handover, cleaning and washing, small-scale work),
- Preventive periodic inspections (without special instruments),
- Control inspections regulated by regulation or law (inspections and other inspections),
- Lubrication (wear analysis, refilling of oils and lubricants, etc.),
- Technical diagnostics (determining the state of the system),
- Preventive replacement of parts,
- Searching for and eliminating weaknesses (innovations),
- Repair and restoration of worn parts of the system,
- Preventive periodic repairs (prophylaxis)—small and medium repairs,
- General preventive repairs with modernization (overhauls),
- Replacement of technical systems with new ones, etc. [5–7].

Maintenance and reliability assessments as well as testing and measuring the reliability of plants with stationary batteries are discussed in detail, and based on this report and data from previous years, Table 2 was made, in which it can be seen that there were no failures of stationary batteries. Areas of focus included: system reliability; system maintenance; condition without failure; consideration of steps that lead to satisfactory reliability through maintenance; review of methodologies for successful reliability assurance; skills that support the success of these methodologies [5].

Table 2. Comparative overview of the state of the system by years. System reliability 2000–2016.

Battery year	2011	2012	2014	2015	2016
Voltage 22 V command voltage	No failure	No failure	No failure	No failure	No failure
Voltage 48 V signaling	No failure	No failure	No failure	No failure	No failure
Voltage $\pm$ 24 V protection system	No failure	No failure	No failure	No failure	No failure

Source [2]

The paper presents the application of the following preventive maintenance procedures on batteries: preventive periodic inspections, preventive periodic replacements, search and elimination of weak points, technical diagnostics, and control inspections. Based on the tests and the results shown in Table 2, it can be concluded that there were no failures on the batteries on which preventive maintenance procedures were applied. We can conclude that in this way there was an increase in the reliability of batteries and thus proved the hypothesis that *preventive maintenance of rechargeable batteries reduces downtime, and thus the degree of safety of thermal power systems.*

6 Conclusion

From the analysis of the test, we can conclude that the preventive maintenance of rechargeable batteries reduces the number of downtimes, and thus the degree of safety of thermal power systems. Rechargeable batteries are the most important part of uninterruptible power supply systems. It depends on them whether the uninterrupted operation of the system will be ensured and how long it will last. In that sense, one of the ways to ensure uninterrupted operation of rechargeable batteries is, among other things, the application of preventive maintenance of rechargeable batteries. This increases the downtime of rechargeable batteries as well as their lifespan.

Each project begins with a report on the current situation and opportunities for improvement. Before attempting to obtain a reliability value, it is very important to do a budget assessment of the maintenance process and understand what the organization is trying to achieve with the reliability of the equipment. Security systems will not fix the problem, they simply optimize what already exists. Their goal is to increase the reliability of the equipment. The most important things to understand before starting are the capacities needed to get started.

Stationary storage batteries are a powerful tool used to achieve consistent reliability improvements, but organizational readiness for maintenance is always a crucial step in the implementation process. Those who use the equipment will build the foundation of the company based on improving reliability by applying carefully chosen methods in the field of maintenance.

References

1. Despotovic, Z.V.: Uninterruptible Power Supply-UPS, Types, Topologies and Components (2017). https://doi.org/10.13140/RG.2.2.16300.87687
2. Records of rechargeable battery control, part of the project documentation: archive of the Termoelektrane—toplane Zrenjanin (1993)
3. STIHL: Safety Information for STIHL Batteries and Products with Built in Batteries. https://static.stihl.com/security_data_sheet/downloads/si-002-2017-01-56.pdf (visited on 02.01.2020)
4. IEC 60896-11:2002, Stationary lead-acid batteries—Part 11: Vented types—General requirements and methods of tests
5. Adamović, Ž.: Tehnologija održavanja, Tehnički fakultet "Mihajlo Pupin". Zrenjanin (2004)
6. Adamović, Ž.: Tehnička dijagnostika, V izdanje, Beograd, OMO (2001)
7. Skok, S.: J. Selektivnost zaštita u sustavima istosmjernih razvoda. Doktorska disertacija, Sveučilište u Zagrebu, Fakultet elektrotehnike i računarstva, Zagreb (2004)

Reverse Engineering and 3D Printing of the Spinning Reel: From Maintenance to the New Product Design

Josip Vukašinović, Ivan Grgić(✉), Jure Marijić, and Mirko Karakašić

Mechanical Engineering Faculty in Slavonski Brod, University of Slavonski Brod, Trg I. B. Mažuranić 2, 35000 Slavonski Brod, Croatia
igrgic@unisb.hr

Abstract. The intensive use of products such as fishing reels in a wet atmosphere near the rivers, seas, and lakes, brings more often need for their maintenance. The chance of malfunctioning is rising in the case of irregular care. In this case, using the principles of reverse engineering and the fused deposition modelling method for 3D printing of parts, the maintaining process of a technical system such as a spinning reel was researched. The Shimano AX 4000FB fishing reel was completely reassembled. All parts of the assembly were checked and the complexity of the disassembly/assembly process was analysed. Considering this, the redesign of the existing spinning reel model Shimano AX 4000FB was proposed using CATIA V5 software to facilitate maintenance. The redesigned parts were printed using the low-cost Creality Ender 3 3D printer. Some parts of the spinning reel were kept, while most of the parts were 3D printed. The housing of the featured spinning reel now consists of three parts instead of two to facilitate easier disassembly and maintenance approach. The thickness of the housing wall has been increased to ensure better resistance to excessive dynamic loads. The bearing bushes were replaced by two ball bearings. After assembling, functionality was tested in the real environment and no problems were encountered.

Keywords: Bearing · Spinning reel · Housing · Maintenance · Reverse engineering · 3D printing

1 Introduction

Generally, fishing reels are divided into baitcasting reels and spinning reels. In fishing reels for baitcasting, the handle is parallel to the spool axis, and in spinning reels, the handle is perpendicular to the spool axis. The spinning reel is characterised by the fact that it does not cause any problems when casting the fishing tackle because the spool does not rotate. This means that the line cannot get tangled in the spool. Currently, almost all spinning reels use a spur gear system [1]. The reel consists of a housing in which there is a gear mechanism, a rotor with a switch, a spool in which the brake mechanism is located and on which the line is wound, and a drive lever that is turned by hand [2]. The gear mechanism determines how strongly and smoothly a fishing reel

D. Blažević et al. (Eds.): OTO 2022, LNNS 592, pp. 106–117, 2023.
https://doi.org/10.1007/978-3-031-21429-5_10

turns, especially when pulling out the fish. A lesser gear can feel like a wind-up toy, while a high-quality gear feels and acts more professional. The price of a fishing reel is a good indicator of what you can expect from the gear mechanism. Gear materials range from brass, zinc, and aluminium to stainless steel. However, for maximum strength and durability in saltwater conditions, stainless steel drive gears and pinions are often the best choices [3]. The available literature has mainly focused on gears and vibration. Inoue and Kurokawa [4] presented a new method of gear design to solve the problem of vibration and pulley degradation. The same authors reported in [5] on the influence of misalignment and manufacturing errors in spur gear teeth. They discovered a new type of interference resulting from the influence of alignment and manufacturing errors. In their paper [6] they presented the development of a vibration simulator for the evaluation of transmission errors in gear pair engagement. Unfortunately, no research was found that deals with the maintenance of fishing reels. This provides an opportunity to look more closely at the function and functionality of spinning reels, their geometry, and the components of the assembly. The questions are: How difficult is it to disassemble or assemble the spinning reel? What is the accessibility of the parts inside the housing? Can 3D printing and reverse engineering help in case of malfunctions? etc. Answering such questions could help in the maintenance of spinning reels and/or could lead to the new product design, as in this case. A necessary step in reverse engineering or designing new products is to create a CAD model. The use of CAD systems leads to the rapid introduction of new products in the market [7]. By creating the 3D CAD (computer-aided design) of the, e.g. reel housing, for example, in the event of a malfunction, one could simply 3D print it instead of buying a new one or presenting a prototype of a new product based on a commercially available product.

For the purposes of this work, the Shimano AX 400FB spinning reel was considered. The maintenance of the spinning reel provides an opportunity to design a new functional prototype of a product. This is explained in more detail below.

2 Materials and Methods

2.1 Maintenace of the Shimano AX 400FB Spinning Reel

It is important to emphasize that the maintenance procedure was done by the assessment and the free will of the authors. No maintenance procedure was followed or owned by the spinning reel manufacturer.

The spinning reel was removed from the fishing rod and the fishline was rolled out from the line roller. The wet soft rag was used to clean up all of the surfaces from dirt and a visual inspection was carried out to notice possible surface damage and/or cracks. The disassembling procedure started with the housing cap removal (Fig. 1) and ended up at the time at which there were no parts to be removed anymore. It was a physical "exploded" view of all of the spinning reel parts.

As it was done for the housing inspection, the same procedure was done for the internal parts. A universal degreaser was used for any spotted grease inside the reel housing, especially on the gear transmission. After detailed cleaning, the damage was found only on the rotor cam (Fig. 2). For better perception, all consisting parts can be seen in Fig. 3.

Fig. 1. Interior look after the housing cap removal.

Fig. 2. The rotor cam wear.

2.2 Function Analysis of Parts

The function of the reel is to enable the easy and simple extension and shortening of the active part of the line or braid. Long-distance fishing and larger fish require the use of fishing reels. The spool on which the line is wound enables the storage of large quantities of line that is wound on it with the help of a mechanism that moves by hand. The line can be unwound in free form from the spool or in a controlled manner utilizing a mechanism in charge of braking. The skills of the fisherman and the quality of the equipment he uses are important in the fight against fish. During the fight, the fish pulls out the line and bends the rod, after which it gradually gets tired. The brake provides the fisherman with safety during the fight against fish. The brake is required to be reliable, non-slip, and non-blocking. The brake cushions the blows of the fish, and proper use prevents the line from cracking.

The quality of the role depends on its durability when extracting fish. The quality of the roll is higher if the reel has a larger amount of bearings (Fig. 4). The diameter, height, depth, shape of the upper part of the spool, and the size of the spool stroke, which ensures uniform and correct winding of the line, are important. The choice of fishing reel depends on the fishing technique we want to fish. In most cases, the technique is related to the type of fish, so we choose the accessories we will use in proportion.

The spin reel for fishing is a complex mechanism that serves to transmit power and movement. The action of the hand force on the lever creates a moment that causes the main gear to move. The main gear acts on the auxiliary gear which drives the main shaft along which the spool moves vertically. Also, the main gear is engaged with a toothed bushing that causes the rotor to rotate around the spool and line winding.

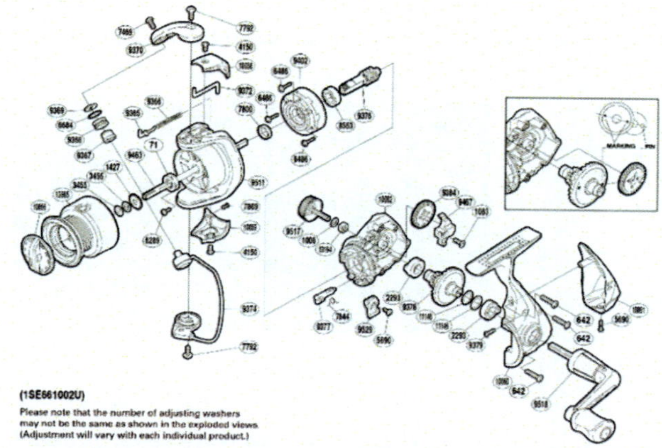

Part No.	Description	Part No.	Description	Part No.	Description
RD 0071	Rotor Nut	RD 7869	Screw	RD 9463	Main Shaft
RD 0642	Screw	RD 8553	Ball Bearing	RD 9467	Oscillating Slider
RD 1008	Handle Lock Washer	RD 9365	Bail Spring Guide	RD 9511	Rotor
RD 1083	Screw	RD 9366	Bail Spring	RD 9517	Handle Screw Cap
RD 1427	Spool Support	RD 9367	Line Roller Bushing	RD 9518	Handle Assembly
RD 2293	Drive Gear Bushing	RD 9368	Line Roller	RD 9529	Anti-Reverse Lever
RD 3455	Spool Washer	RD 9369	Line Roller Washer	RD10056	Drag Knob
RD 4150	Screw	RD 9370	Bail Arm	RD10058	Bail Spring Cover
RD 5690	Screw	RD 9372	Bail Trip Lever	RD10059	Bail Hold Support Guard
RD 6289	Screw	RD 9374	Bail Assembly	RD10060	Side Cover
RD 6486	Screw	RD 9376	Pinion Gear	RD10061	Rear Protector
RD 6684	Line Roller Spacer	RD 9377	Anti-Reverse Cam	RD10062	Body
RD 7469	Screw	RD 9378	Drive Gear	RD11146	Washer
RD 7792	Screw	RD 9379	Screw	RD13194	Handle Screw Lock
RD 7800	Rotor Ring	RD 9384	Oscillating Gear	RD13665	Spool Assembly
RD 7844	Anti-Reverse Cam Spring	RD 9402	Roller Clutch Assembly		

Fig. 3. Shimano AX 4000FB technical documentation [8].

In the reel Shimano AX 4000FB there are two bearings, one ball, and one roller bearing. Both bearings are located on the main axle, and their properties contribute to the smoothness of the transmission. In addition, there are two bearing bushes located on the shaft on which the handle is mounted, which rotates by the force of a human hand.

Fig. 4. Drive gear bearing bush and side cover.

These two bearing bushes (Fig. 5) can provide space for upgrading, i.e. by replacing them with two double-closed ball bearings, the reel will significantly gain in quality. However, the reconstruction of the reel housing is required to enable such implementation.

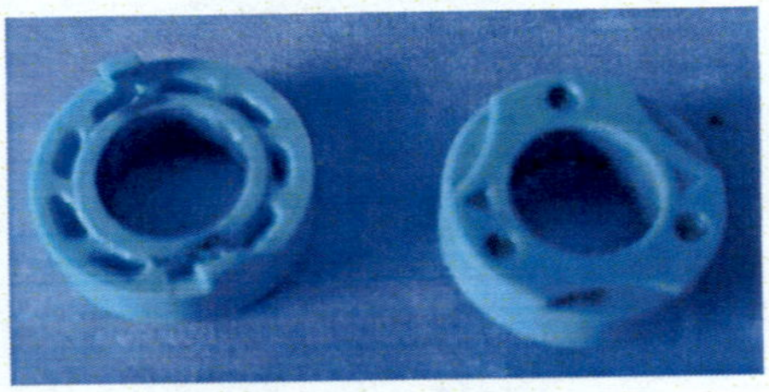

Fig. 5. Bearing bushes.

Furthermore, according to Fig. 1, all internal parts are located on one side of the reel housing. If someone wants to reach the component that is positioned deep inside the housing and check it for malfunction, it has to remove all other parts positioned above it. Again, here is noticed the possibility for the reconstruction of the housing to rise the maintenance efficiency.

Additionally, the wall thicknesses of the housing will be increased, which additionally contributes to the durability and load-bearing capacity of the entire mechanism, and the rod handle will be strengthened to ensure the connection between the reel and the rod. The housing will have three parts instead of two. If the rod handle breaks, it will be possible to print only that part and replace it. Also, with prolonged use, the cam on the rotor wears out, which serves to ensure the immobility of the rotor, but this will no longer be a problem because it will be possible to 3D print out the complete rotor and replace it. Moreover, to implement the industrial design, the reel handle will get some retro style.

2.3 Redesign of the Spinning Reel

Dassault Systemes Software CATIA V5 [9] was used for the reconstruction procedure.

Reel Housing and Cap. The goal is to achieve housing (Fig. 6) with the possibility of easier installation, and replacement of one of its parts after possible fatigue. Also, the new shape of the housing will provide the possibility of installing two additional bearings that make the transmission smoother. All of the housing parts will be connected with four M3 screws. The design of the housing cap was adjusted to this new reel housing and the correlation to the original can be seen in Fig. 7.

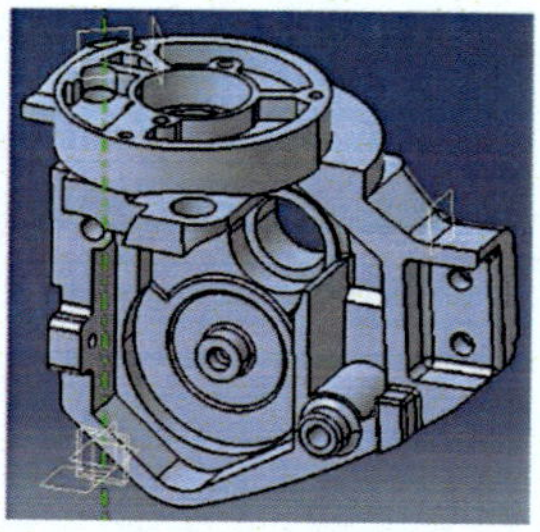

Fig. 6. The new reel housing.

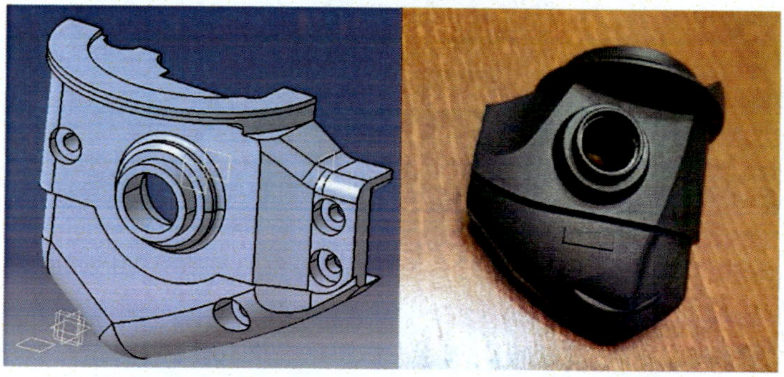

Fig. 7. Redesign and the original shape of the cap.

Rod Handle. Improper storage and careless handling of the reels often cause them to fall to the ground. Due to its construction, in most cases, the reel falls on the rod handle and breaks. In almost all reels, the rod handle and the housing form one part, and thus forms an inseparable joint, if the rod handle breaks, it is impossible to repair the damage. The design of a new independent rod handle (Fig. 8) will allow it to be easy to change in case of damage regardless of the housing.

Housing Protector. The housing protector (Fig. 9) is located on the lower part of the housing as protection against shocks and water penetration into the interior. If the reels are handled, stored, and maintained properly, they can have an unlimited lifetime, but it often happens that fishermen inadvertently place the rod together with the reel on the ground, which can damage the housing. To avoid such damage, a housing protector was constructed that fits perfectly with this new housing design.

The Rotor of the Reel. By rotating around the spool it allows the fishline to be wound. At the other end of the rotor, there is a space provided for the lever and for the spring

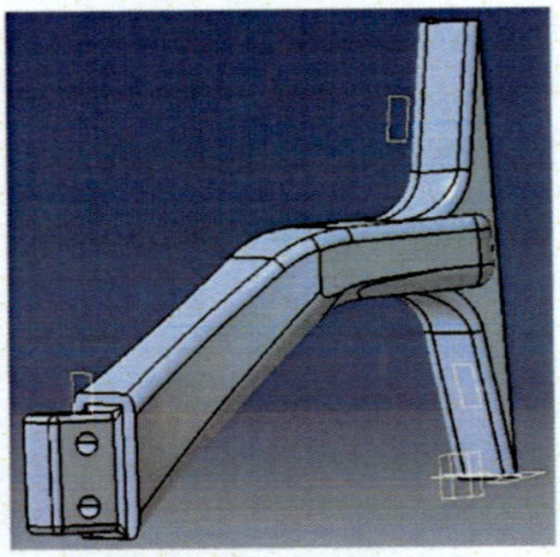

Fig. 8. The rod handle.

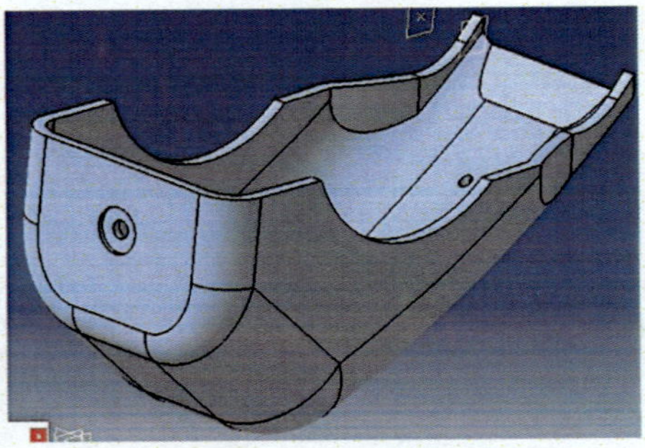

Fig. 9. The 3D model of the housing protector.

which returns the switch to its initial state (Fig. 10). To protect these parts and their proper functioning, a rotor side-cover was designed (Fig. 11), which is fastened with two screws to the rotor.

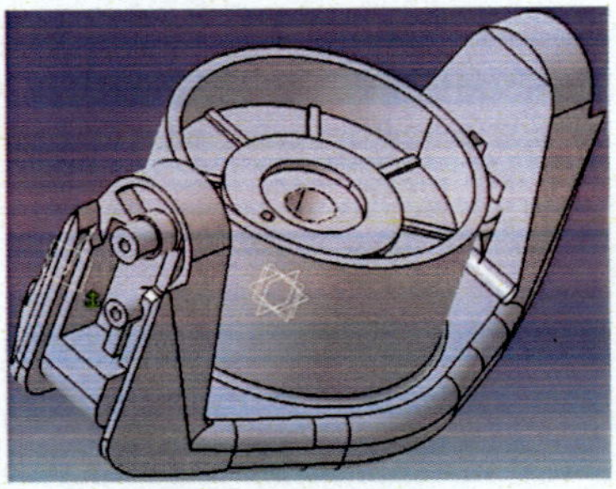

Fig. 10. The rotor of the reel.

The final 3D model of the assembly can be seen in Fig. 12.

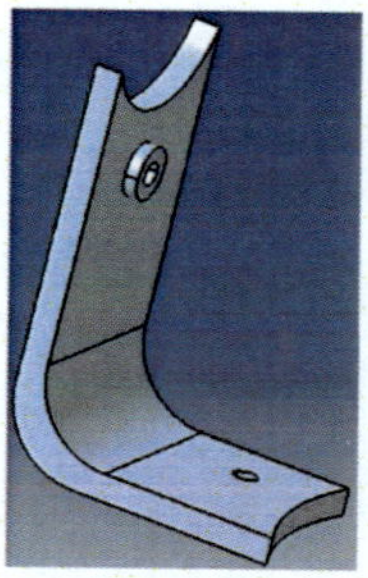

Fig. 11. The side cover of the rotor.

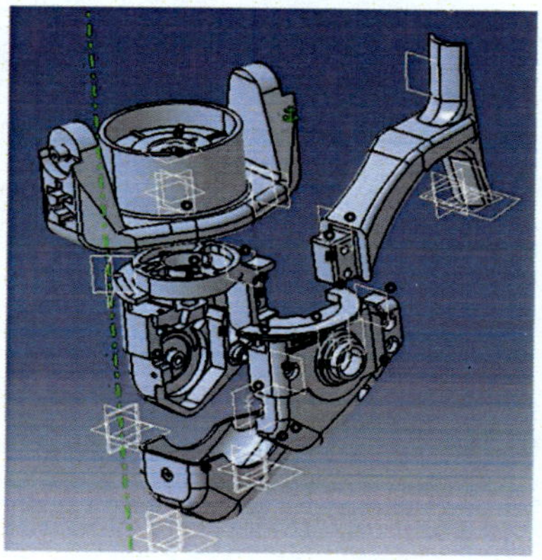

Fig. 12. Exploded view of the spinning reel.

2.4 3D Printing

For 3D printing (Fig. 13), the Ultimaker CURA slicing software was used. All parts were saved as the STL file and imported into the slicing software. The low-cost Creality Ender 3 3D printer was used. The parameter of the 3D printing can be seen in Table 1.

Table 1. 3D printing parameters.

Parameter	Value
Temperature	205 °C
Base temperature	60 °C
Infill	100%
Speed	30 mm/s
Material	PLA
Layer height	0.12 mm

Fig. 13. Finished 3D printing of the housing cap.

3 The Assembly of the New Spinning Reel

Before mounting or assembling the entire system, it is necessary to lubricate the bearings and gears, and other moving parts. Reell lubrication is possible with oil or grease depending on the construction and the possibility of lubrication. "Penn Fishbowl Grease" was chosen for lubrication, i.e. white grease that is applied to all moving parts and guides.

After inserting all of the moving parts into the housing (Fig. 14) and closing the housing with the cap, these parts are connected to each other and to the handrail using four M3 screws that form a detachable connection with the nuts (Fig. 15).

Then, a bushing and a roller bearing with a system that prevents or rotates the rotor in the opposite direction via a switch are pulled onto the upper part through the shaft. After the lower part is assembled into a compact unit, assembling the rotor parts was the next step. The assembled rotor is connected to the housing using the main M6 nut. The housing and rotor now form a compact unit. The spool with the braking system handles and housing cap installation was the last step (Fig. 16).

The final assembly of the new design of the spinning reel is presented in Fig. 17.

Sine, the following the maintenance procedure of the original Shimano AX 4000FB was not followed or either provided, the maintenance of the proposed spinning reel is suggested below.

4 Maintenance Protocol of the New Spinning Reel

To effectively perform maintenance, first, the appropriate tool has to be chosen to disassemble the roll and separate all moving parts. Then the appropriate cleaner and lubricant

Fig. 14. Moving parts in the reel housing.

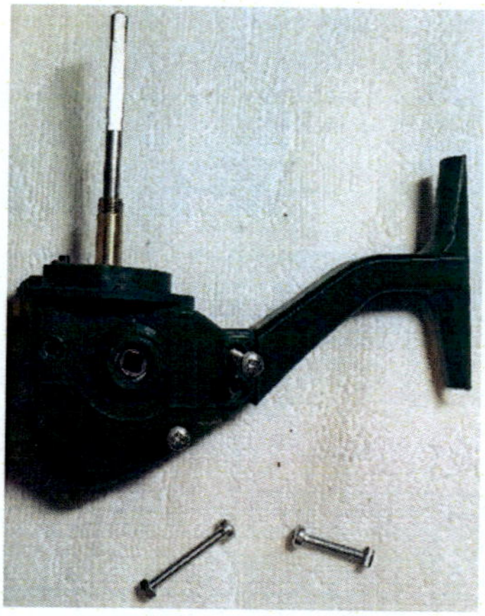

Fig. 15. Housing connection with the rod handle.

to remove deposits of old lubricant and impurities from the environment that have entered the mechanism (sand, dust,..) have to be chosen. Deposits of old lubricant and dirt cause gear wear, tooth gap, and difficult transfer of rotation from the lever to the rotor.

Maintenace protocol is as follows:

1. Remove the cover holding the spool on the shaft.
2. Remove the fuse holding the brake discs in place to clean old grease deposits and dirt caused by external influences.
3. Remove the lever and check the rotation of the bushing around the shaft.
4. Remove the bolt holding the main nut to allow the rotor to separate from the roll housing. By removing the rotor, accumulations of old grease on the main shaft can be noticed and removed.

Fig. 16. Spool installation.

Fig. 17. The new spinning reel design.

5. By opening the rotor cover, check the spring used to open the switch, and check the lever on the housing cam to prevent the rotor from turning when the hook is thrown.
6. Remove the three retaining bolts holding the anti-reverse roller bearing system and the ball bearing on the main shaft to allow the removal of the main bolts used to secure the housing.
7. Check the correctness of the main bearing and lubricate if necessary.

8. Remove the two screws that secure the housing cover, then loosen the four main screws that secure the housing and the handle. The internal parts are visible and can be cleaned.
9. Check the functionality of the two closed ball bearings and lubricate them if necessary.
10. Remove all remaining moving parts (gears, shafts,..), clean them with the selected cleaning agent, and lubricate them.
11. Upon completion, assemble in reverse order.

5 Conclusions

The use of reversible engineering and additive technologies in the design and construction of a redesigned engineering system such as a spinning reel for fishing has enabled the development of a prototype for a new product. All modelled parts were successfully printed and assembled with an improved mechanism of a redesigned reel into a functional product. The problem was the complex gearbox inside the fishing reel, which had to be precisely fitted into the redesigned housing. Disadvantages in terms of 3D printing speed and poorer surface quality require additional machining of the material. It is expected that these shortcomings will be eliminated in the future.

More importantly, the design process also resulted in a maintenance protocol.

References

1. LNCS Homepage. https://doi.org/10.1007/978-3-319-70802-7_8. Last accessed 14 March 2022
2. NA UDICI Homepage. https://naudici.com/role-i-multiplikatori-2/. Last accessed 14 March 2022
3. SPORT FISHING Homepage. https://www.sportfishingmag.com/fishing-reel-gears-explained-0/. Last accessed 15 March 2022
4. Inoue, T., Kurokawa, S.: Derivation of path of contact and tooth flank modification by minimizing transmission error on face gear. J. Adv. Mech. Des. Syst. Manuf. **6**(1), 15–22 (2012)
5. Inoue, T., Kurokawa, S.: Influence of manufacturing error on transmission error of a face gear with controlled curve and proposal of improved manufacturing process. In: International Conference on Gears, pp. 1043–1054. VDI Wissensforum GmbH, Munich (2013)
6. Inoue, T., Kurokawa, S.: Development of vibration simulator for transmission error evaluation of gear pair engagement. In: The 14th Machine Design and Tribology Division Meeting in JSME, pp. 151–154. Nagano (2014)
7. Havrylenko, Y., Cortez, J.I., Kholodniak, Y., Alieksieieva, H., Garcia, G.T.: Modelling of surfaces of engineering products on the basis of array of points. Tech. Gazette **27**(6), 2034–2043 (2020)
8. SHIMANO Homepage. https://dassets.shimano.com/content/dam/global/cg1SHIFSPL/EV/06AX1000FB.pdf. Last accessed 16 March 2022
9. CATIA V5 Homepage. https://www.3ds.com/products-services/catia/. Last accessed 17 March 2022

Development of V-Twin Motorcycle Valve Stem Seal Non-invasive Removal Tool

Josip Cumin(✉), Daniel Novoselović, Dejan Marić, and Tomislav Šolić

Mechanical Engineering Faculty in Slavonski Brod, University of Slavonski Brod, Slavonski Brod, Croatia
jcumin@unisb.hr

Abstract. Vehicle maintenance over the last few years has become more expensive due to the COVID related crisis, spare parts transportation problems, and latest price rises - related to war over Ukraine. There is also evident rise in fuel and lubricant prices, which is related to crude oil price rises. So, due to this reasons, older types of motorcycles and automobiles are often serviced in small repair shops, or maintenance is done by owners instead of certified dealer shops. This paper deals with tool construction designed for changing valve stem seals in V-twin motorcycle, with special emphasis on non-invasive procedure. With this tool, there is no need for head removal (as it is done in conventional procedure), thus leading to lower maintenance costs. Top-end gaskets sets are expensive, and work related to heads removal of V-twin motorcycle engine is time consuming. For the design of this oil seal removal tool, a reverse engineering technology is used, such as CAD (computer aided design, CAM (computer aided machining) and FEM (finite element analysis).

Keywords: CAD · CAM · FEM · Motorcycle · Maintenance · Tool

1 Introduction

Vehicle maintenance is very important factor in automotive and motorcycle reliability. Special maintenance procedures are related to other types of transportation devices like marine ships, boats, jet skis, aircraft piston-type and turbine type engines.

When these two types of maintenance procedures are compared, it can be said that automotive maintenance is less cumbersome and more user friendly, due to the robustness of automotive ICE (*Internal Combustion Engine*) engines, and lesser criteria for engine maintenance points.

It is well known fact that the engines (machines) which can endanger lives of people due to the failure, have to pass rigorous inspections, and on-board systems of such engines are often redundant for the same reason (for example ROTAX 912 ULS/S which has dual electronic ignition, and dual carburetors [1]).

So, in today's society, many ICE engines from automotive area have maintenance in "Do It Yourself" variants, due to the several reasons [2].

D. Blažević et al. (Eds.): OTO 2022, LNNS 592, pp. 118–132, 2023.
https://doi.org/10.1007/978-3-031-21429-5_11

Main ones are:

- During 2021 Covid outbreak car mechanic waiting lists were extremely long
- Maintenance parts and tools were sparse (due to the lockdowns)
- Thanks to global network, maintenance knowledge is very easily obtainable through on-line user and repair manuals
- Through forums, the users often share experience, which is very important

In recent months of year 2022, the prices of crude oil and derivates has skyrocketed, due to the Russia-Ukraine war. For this reason, the maintenance of transportation vehicles has become also very expensive.

2 Description of Problem

This paper deals with V-twin, 75°, 950 cm^3 motorcycle which has had a history of oil consumption. In this case study, based on several former maintenance inspections, it has been noticed that engine oil has been present on the intake valves. The oil on the intake valves can be present due to the either worn valve stem guides, or damaged valve stem seals (burned out valve stem seals due to high temperatures in high compression ratio engines).

In order to replace valve stem seals, it is necessary to remove carburetors, intake manifold, exhaust, cooling system and electrical system. On some engines, only the cylinder head(s) can be removed, while on other configurations the whole engine needs to be taken out of the vehicle and taken on the stand.

Meanwhile, there is known procedure for replacement of only valve stem seals while engine is not disassembled, and which had been utilized in this paper for maintenance of the fore mentioned engine.

The benefits are lower overall repair costs, no need to change oil and oil filters, no need for engine removal from chassis. For comparison, Table 1 shows bill of materials and price estimate for the conventional engine disassembly.

Table 1. Estimate of bill of materials for the repair.

No	Item	Quantity/Piece	Price/Eur
1	Valve-stem sealing racing	8	3.05·8 = 24.4
2	Cylinder head gasket 0,5 101mm	2	2·46.75 = 93.5
3	Exhaust pipe gasket	2	2·11.69 = 23.38
4	Copper self locking nut	4	4·1.73 = 6.92
5	Oil filter	1	13.01
6	Oil 10W60	4 L	3·13.29 = 39.87
7	Spark plugs CR8EK	2	2·10 = 20
8	Coolant	2 L	2·8.24 = 16.48
Total:			237.56 €

After engine top end disassembly there is possibility of oil contamination, so it is best practice to change oil and oil filter after engine top end overhaul. Oil can also be tested with different methods [3–6], and based on the results it can be reused if it is of a good quality, but in such small quantity - the best bet is using fresh oil after overhaul procedure where cylinder heads are removed.

On the non-invasive valve stem seal approach, from Table 1 – the cost is calculated by adding items No. 1., No. 7., No. 8., which leads to total parts cost of 60.88 €. The tool proposed here, has been manufactured from metal scraps and easily obtainable hardware-store elements. The tool production cost should be calculated and added to the total sum, but since the tool was manually produced from scraps, the tool price could not be exactly determined.

3 Tool Design and Calculations

Figure 1 shows cylinder head with camshafts removed (left), and on the right side that same cylinder head transferred to AUTO CAD, and made as dxf-file. This was necessary for two reasons, first – the making of 3D model for the base plate of tool; and second for the purpose of generating G-code in CAM software.

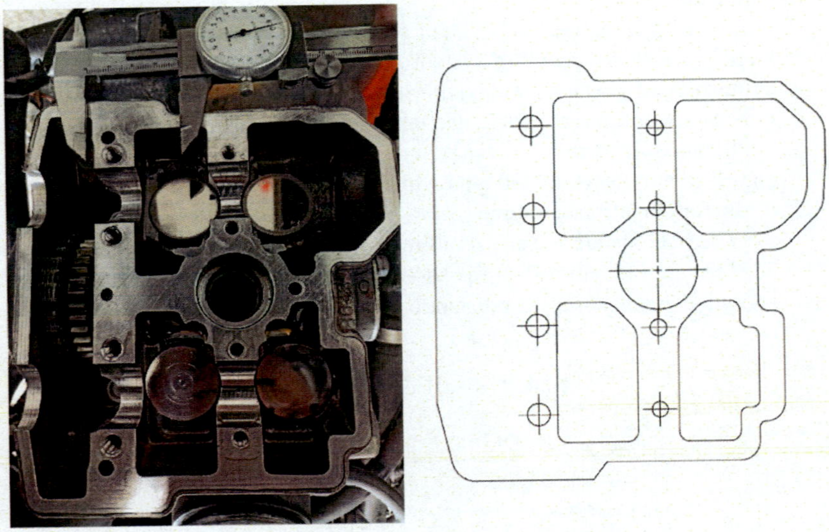

Fig. 1. Cylinder head with cams removed.

Figure 2 shows the proposed tool assembly made in Autodesk Inventor educational edition.

For the FEM analysis, the forces acting on the valve spring should have been estimated first. Measuring of the spring characteristic k has not been an option, since the tool was required in the first place to remove springs.

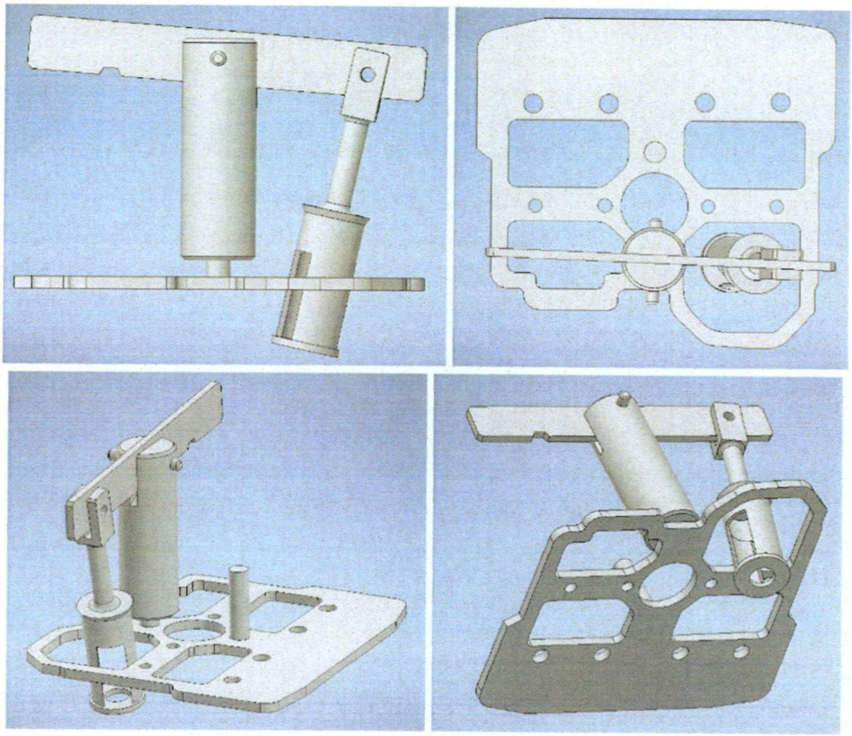

Fig. 2. Valve stem seal removal tool 3D model assembly (Autodesk Inventor).

For this purpose, the technical data from repair manual has been used to calculate maximal spring force based on the assumption that the spring has been made from the strongest spring material available. In this way the calculated spring force would have been overestimated (which is in favor of tool safety factor).

Figure 3 shows outer valve with 29 mm outer dimension, 43 mm unloaded length, 3,7 mm spring wire thickness and with 4 active coils. The inner (safety) spring can be seen also, and the dimensions of inner spring are 20 mm of outside diameter, 40 mm unloaded (loose) spring length, 2.5 mm wire diameter and 5.5 active coils.

Figures 4 and 5 shows Autodesk Inventor spring calculator for the outer (main) valve spring. It can be seen that the maximal spring force is equal to the $F \approx 493$ N.

Figure 6 show calculated data for inner spring, showing maximal spring force of $F = 234$ N.

Total valve spring loading force can be calculated as:

$$F_u = F_o + F_i = 492.4 + 234 = 726.4\,\text{N} \tag{1}$$

In order to account for the unknowns, this force is enlarged 30%; that is:

$$F_{tot} = 1.3 \cdot F_u = 945.1 \approx 950\,\text{N} \tag{2}$$

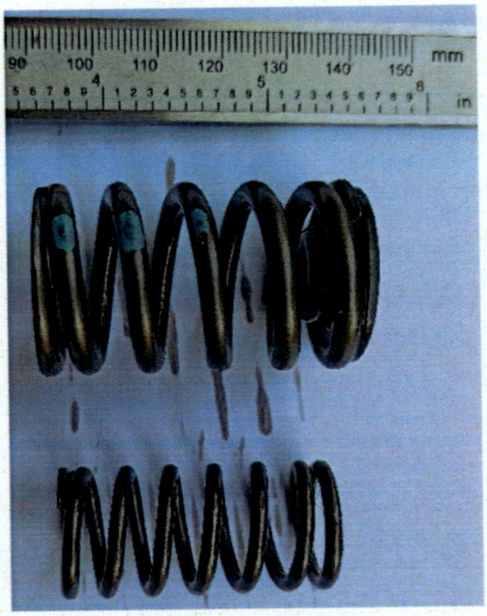

Fig. 3. Valve springs, outer (main), and inner (safety).

Figure 7 shows schematic for loads on the spring compression mechanism.

For the criterion $\sum F_x = 0$:

$$- F_1 \cdot \sin 11^\circ + F_{AX} = 0$$
$$F_{AX} = F_1 \cdot \sin 11^\circ = F_{tot} \cdot \sin 11^\circ = 181.3\text{N} \quad (3)$$

For the equilibrium criterion $\sum M_A = 0$ it can be written:

$$- F_1 \cdot a + F_L \cdot b = 0$$
$$F_L = \frac{F_1 \cdot a}{b}, \text{N} \quad (4)$$

For the equilibrium criterion $\sum M_B = 0$ it can be written:

$$- F_1 \cdot \cos\left(11^\circ\right) \cdot (a + b) + F_{AY} \cdot b = 0$$
$$F_{AY} = \frac{F_1 \cdot \cos(11^\circ) \cdot (a + b)}{b}, \text{N} \quad (5)$$

From Fig. 7., it can be seen that there are two center holes, and by changing their position, the distance of *a*, and *b,* and their ratio also changes – which leads to different loading forces. This is shown in Table 2.

These loads are further used in the FEM analysis.

4 Finite Element Modeling (FEM)

Previously, the loading forces have been determined. For this purpose, a simple equations have been used (1–5).

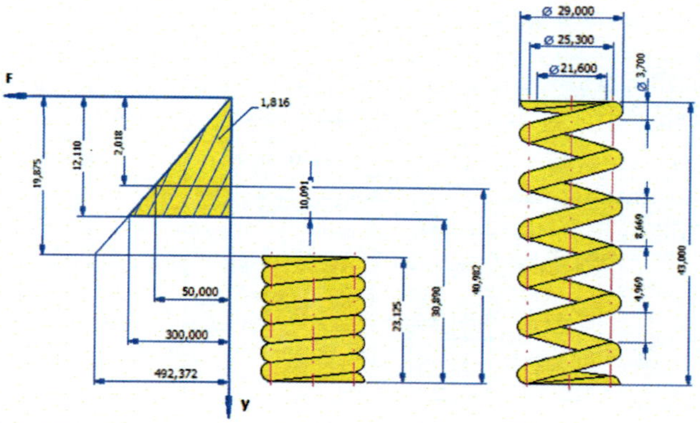

Results

Space between Coils of Free Spring	a	4,969 mm
Pitch of Free Spring	t	8,669 mm
Stress Concentration Factor	K_W	1,000 ul
Spring Constant	k	24,773 N/mm
Min. Load Spring Deflection	s_1	2,018 mm
Total Spring Deflection	s_8	12,110 mm
Limit Spring Deflection	s_9	19,875 mm
Limit Test Length of Spring	L_{minf}	26,074 mm
Theoretic Limit Length of Spring	L_9	23,125 mm
Spring Limit Force	F_9	492,372 N
Min. Load Stress	τ_1	63,595 MPa
Max. Load Stress	τ_8	381,572 MPa
Solid Length Stress	τ_9	626,251 MPa
Critical Speed of Spring	v	7,461 mps
Natural Frequency of Spring Surge	f	480,415 Hz
Deformation Energy	W_8	1,816 J
Wire Length	l	526,240 mm
Spring Mass	m	0,044 kg
Spring Check Result		Positive

Fig. 4. Autodesk Inventor spring calculator for the outer (main) valve spring.

Fig. 5. Outer valve spring calculated data (Autodesk Inventor).

The proposed base plate has had multiple constraints (loading conditions), thus a finite element method has been used to calculate displacements, stresses and strains since it can approximate results very well. The Autodesk Inventor educational edition has been used for the numerical calculation, since it is conveniently integrated with CAD/3D

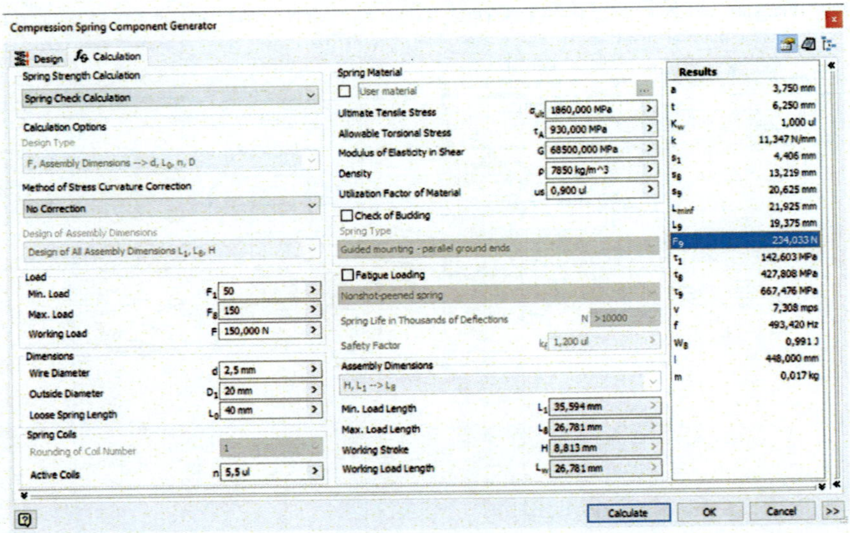

Fig. 6. Inner valve spring calculated data (Autodesk Inventor).

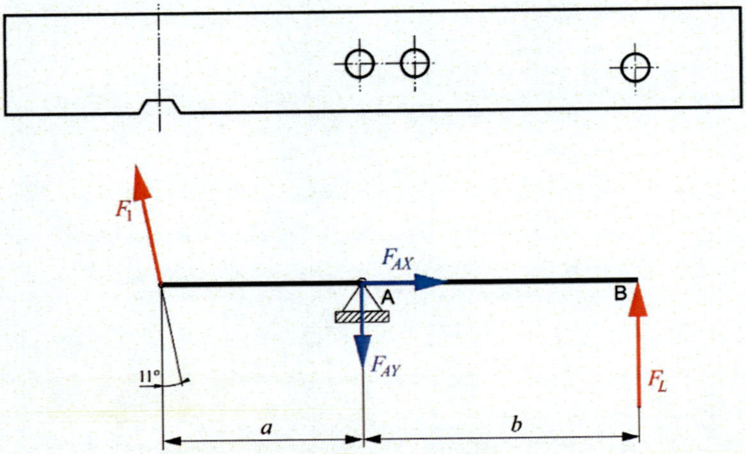

Fig. 7. Schematic of loads on a spring compression mechanism beam.

modeling package, and it has been very well proven in the linear elastic numerical analysis [7–11].

For the base plate material, the S355J2+N (1.0570/St 52-3N) steel has been chosen. Table 3 shows material chemical composition, and typical mechanical properties are given in Table 4.

After assembly of only critical loading elements (shown in Fig. 8), the boundary conditions have been applied. As shown, the eight holes on base plate have been fully

Table 2. Calculated loads on the spring compression mechanism beam.

Lever length/mm	F_{AX}/N	F_{AY}/N	F_L/N
$a = 60$ $b = 43.5$	181.3	2218.8	1286.3
$a = 48$ $b = 55.5$	181.3	1739.1	806.5

Table 3. Chemical composition of S355J2+N steel [12].

C/%	Si/%	Mn/%	Ni/%	P/%	S/%	Cr/%	Mo/%	Al/%	Cr + Mo + Ni
0.22	0.55	1.6	0.3	0.035	0.035	0.3	0.08	0.02	<0.048

Table 4. S355J2+N mechanical properties [13].

$R_{p0,2}$/MPa Yield strength	R_m/MPa Tensile strength	A_5/% elongation	HB	Impact strength
355	470–630	22	140–190	27J (−20 °C)

constrained. Bracket support on the top side of base plate has been set as glued (no motion or friction at the place of contact). At the upper-bracket hole, the force F_l has been applied in the vertical y-axis direction.

At the carrier of the spring compression beam, both x-axis and y-axis loads have been applied.

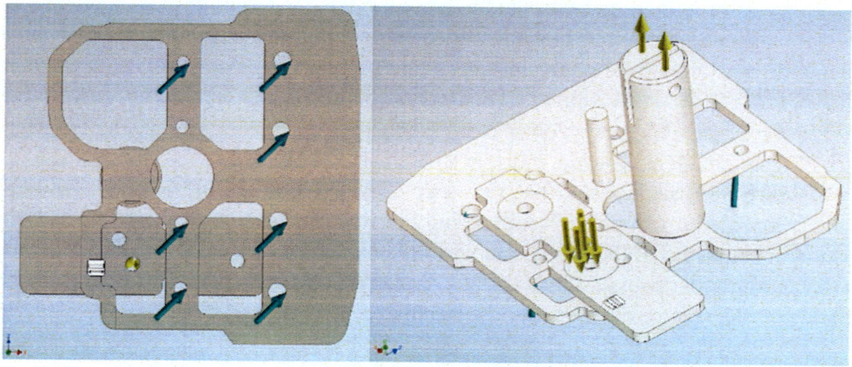

Fig. 8. FEM model with boundary conditions and loads.

Figure 9 shows Von Mises stress in the assembly for the initial tool configuration.

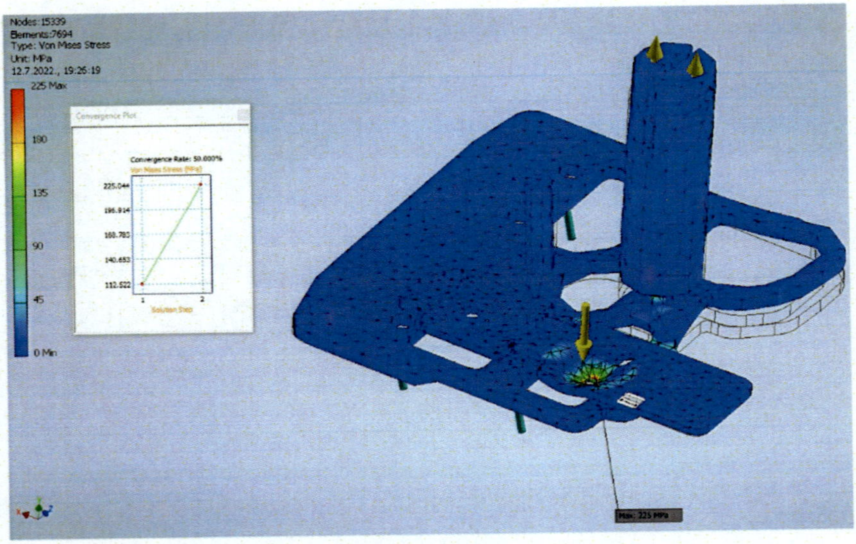

Fig. 9. Initial Von Mises stresses.

It can be seen that the top bracket has had large stresses located near the hole end. This is due to the poor force definition in FEM analysis. For the next step an M8 universal washer has been set by splitting top-bracket surface due to the M8 washer dimensions (outer diameter of 24 mm). On this surface the corresponding pressure has been applied for the all subsequent numerical simulations.

Furthermore, on the surfaces with higher stress gradient, a refined mesh has been used. Average refined mesh element size has been set as 1 mm. This can be seen on Figs. 10, 11, and 12.

Figure 10 shows top view of the base and bracket assembly with refined mesh visible.

Figure 11 shows base and bracket top view without mesh with visible place for M8 washer at the upper bracket – located at right side in the image. Four areas with the highest amount of equivalent stress are highlighted, where largest stress is indicated as $\sigma = 191{,}4$ MPa (same as in Fig. 10).

Figure 12 shows lower side of the base plate, with indicated 1st principal stress (maximal tensile stress induced in the part due to the loading/shear stress is considered zero). It can be seen that the largest tensile stress value is $\sigma = 214.8$ MPa. When considered guaranteed yield strength of material (Table 4; $R_{p0,2} = 355$ MPa) it can be concluded that the base plate will not enter plasticity area, and material can reversibly endure that stress.

The factor of safety can be calculated:

$$S = \frac{R_{p0,2}}{\sigma} = \frac{355}{214.8} = 1.65 \tag{6}$$

There is question of tool rigidity at the totally stressed tool. The FEM displacement in the y-axis direction is shown in the Fig. 13. Total displacement at the most distant

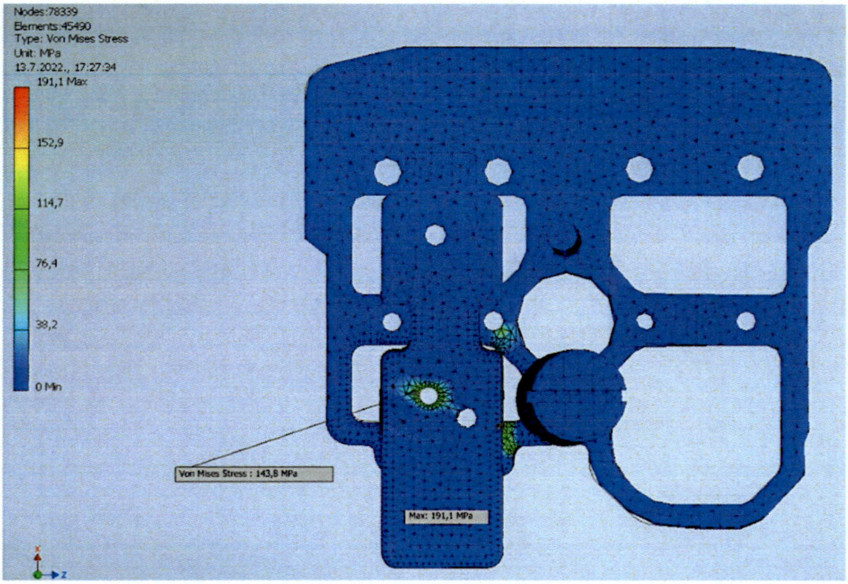

Fig. 10. Top view of the base and bracket with refined mesh visible.

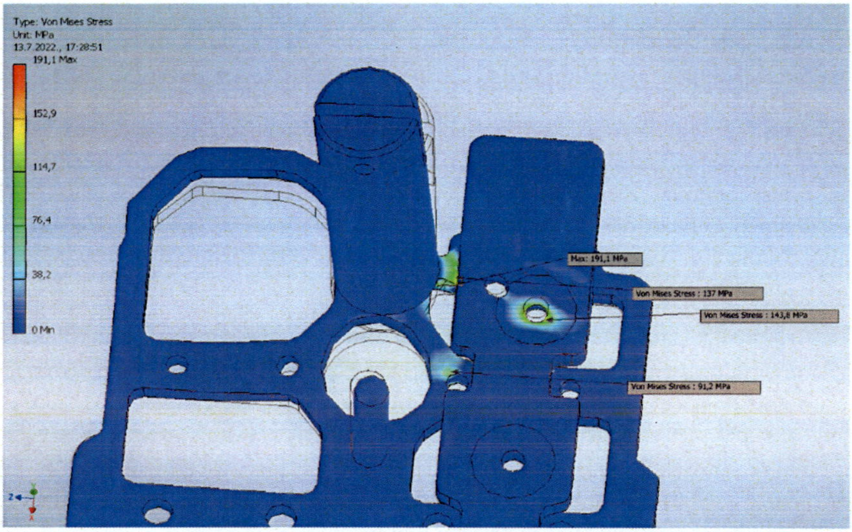

Fig. 11. Base and bracket top view without mesh (visible place for M8 washer at the upper bracket – located right in the image).

part of the base plate us $u_y = 0.04814$ mm, which is reverible during loading/unloading cycle. This amount of displacement is considered irrelevant for the tool functioning, and

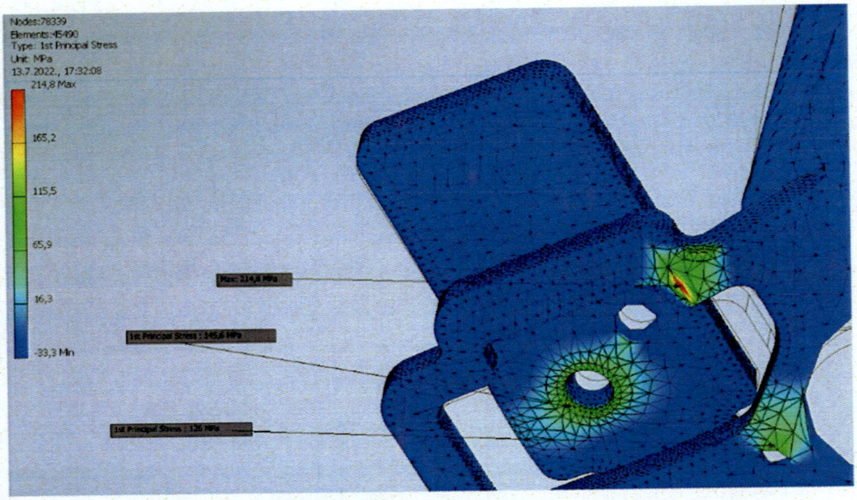

Fig. 12. The place of 1st principal stress (tensile) in the base plate.

it can be concluded that the base plate is properly dimensioned with the given geometry and plate thickness of 5 mm.

5 Application

Figure 14 shows valve spring compression tool application. Base plate has been mounted on the cylinder head, and bolted with four M6x25 screws, and four M8x25 screws. Between cylinder head and base plate of the tool, paper gasket has been cut to the shape and mounted. Paper gasket has been used not to damage surface of the cylinder head since cylinder head and camshaft bracket are matching parts (they are assembled and torqued to the specs, and then machined to obtain required tolerances). So, it has been imperative not to damage the matching surfaces, and not to over tighten the tool screw.

Figure 15 shows removal of valve collet halves after springs have been compressed, and on the right side - removed valves and valve spring retainer plate.

Figure 15 shows dismantled springs, round valve spring retainer plate and valve collet halves.

Figure 16 shows valve stem seal which has been removed and replaced with the new one. The rubber of old valve stem seal has been found hardened in respect to the new part. This is due to the high operating temperatures of V-twin cylinders with high compression ratio and high-power output.

Figure 17 shows final operations of assembly and adjustment of valve clearance with shims and tappets (marked 1–4).

After the maintenance has been finished, it has been found out (by means of OTG, usb endoscope camera), that there has not been evidence of oil residue on the intake valves. This has confirmed practicality of the tool, since maintenance process has been done with same efficiency and lower costs than regular maintenance procedure.

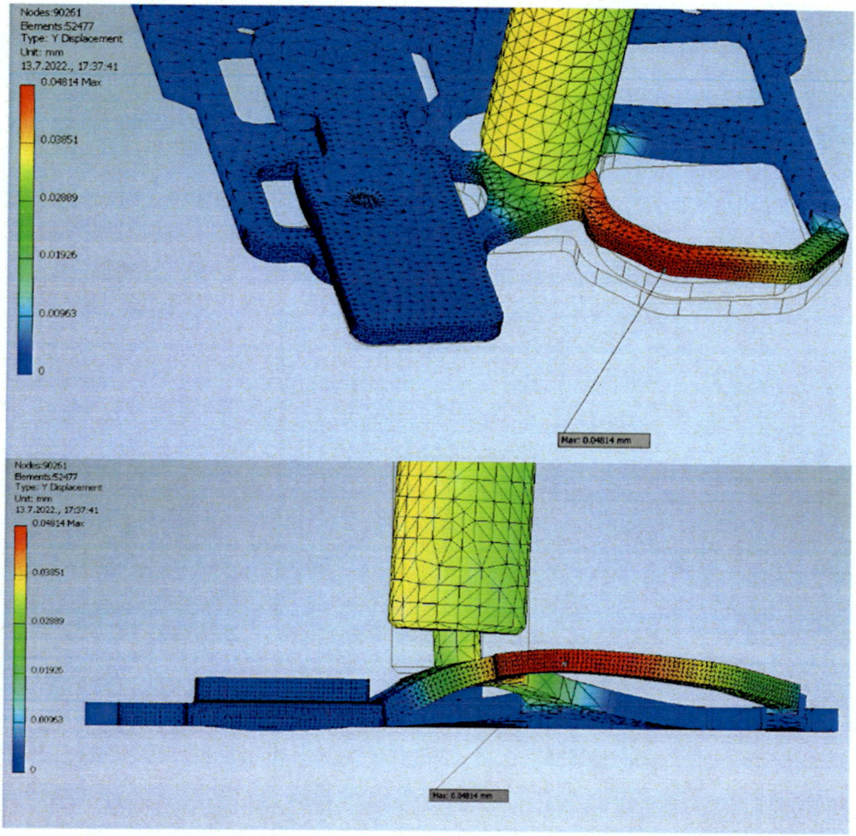

Fig. 13. Maximal displacement u_y (in y - axis direction) of the base plate under loading.

It should be noted that this is a high-risk operation and the authors hold no liability for the damage done with application of this tool.

6 Conclusion

This paper deals with the problem of leaking valve stem seals on the V-twin motorcycle engine. For this repairing in conventional way it is necessary to remove exhaust, carburetors, all cables, camshaft chain and in the end removing of cylinder head itself.

By adding price of top end gasket set, price of oil with filters alongside with large amount of work hours, it can be seen that this conventional maintenance generates high cost for the customer.

In this paper, a different approach has been proposed. The usage of spring compression tool which mounts on the cylinder head can be used to compress valve springs. With the use of compressed air in the cylinder area, the valves are held closed shut. In this way, a valve collet can be removed, and after removal of springs the old valve stem seals

Fig. 14. Valve spring compression base plate (left) and whole tool (right).

Fig. 15. Removal of valve collet halves (left), removed springs and spring retainer plate (right).

can be removed and changed with the new ones. This approach is not factory certified, and has it's own risks, but results of this paper show that the same maintenance job has been done with the only fraction of the cost.

Valve spring compression tool has been designed and calculated by the means of basic mechanics, and with the usage of finite element method. Numerical simulations and construction have been done in Autodesk Inventor educational edition.

Fig. 16. Dismantled springs, round valve spring retainer plate and vale collet halves.

Fig. 17. Finalizing of maintenance procedure – setting of valve clearance.

References

1. Rotax: Rotax powertrains: Rotax 912 S/ULS, URL: https://www.rotax.com/en/products/rotax-powertrains/details/rotax-912-s-uls.html, 4 June 2022
2. Glavaš, H., Karakašić, M., Kljajin, M., Desnica, E.: Essential preventive automobile maintenance during a pandemic. Tech. Gaz. **28**(6), 2190–2199 (2021). https://doi.org/10.17559/TV-20201229130413
3. Zöldy, M.: Engine oil test method development. Tech. Gaz. **28**(3), 1012–1016 (2021). https://doi.org/10.17559/TV-20200122150623

4. Tormos, B., Olmeda, P., Gómez, Y., Galar, D: Monitoring and analysing oil condition to generate maintenance savings: a case study in a CNG engine powered urban transport fleet. Insight **55**(2), 84–87 (2013). https://doi.org/10.1784/insi.2012.55.2.84
5. Sgroi, M.F., Asti, M., Gili, F., Deorsola, F.A., Bensaid, S., Fino, D., Kraft, G., Garcia, I., Dassenoy, F: Engine bench and road testing of an engine oil containing MoS2 particles as nano-additive for friction reduction. Tribol. Int. **105**, 317–325 (2017). https://doi.org/10.1016/j.triboint.2016.10.013
6. Sheng, S.: Monitoring of wind turbine gearbox condition through oil and wear debris analysis: a full-scale testing perspective. Tribol. Trans. **59**(1), 00–00 (2015). https://doi.org/10.1080/10402004.2015.1055621
7. Cipto, Rahangmetan, K.A., Sariman, F., Hariyanto, Bahar, S.B., Sapsal, M.T.: Stress analysis on the frame of peanut skin peeler using autodesk inventor software. In: IOP Conferences Series: Earth and Environmental Science, p. 343 (2019). https://doi.org/10.1088/1755-1315/343/1/012164
8. Weia, Q., Xuanb, L.: Stress analysis of the submarine framework based on inventor. Appl. Mech. Mater. **501–504**, 607–610 (2014). https://doi.org/10.4028/www.scientific.net/AMM.501-504.607
9. Mereuta, V.: Finite element analysis of camshaft using inventor software. Int. J. Res. Appl. Sci. Eng. Technol. **9**(XII), 906–912 (2021). https://doi.org/10.22214/ijraset.2021.39319
10. Suryanarayana, S.M.: Static model and analysis of micro gas turbine using autodesk inventor and autodesk nastran software. IOSR J. Mech. Civ. Eng. (IOSR-JMCE) **15**(4), 18–30 (2018)
11. Sakutukwa, M.Z., Jain, A.: Finite element approach to stress analysis for failure prediction of a six DOF hybrid kinematic machine for on-site machining. Int. J. Sci. Res. **5**(4), 2340–2344 (2016)
12. European Steel and Alloy Grades.: S355J2(+N), 1.0570(dubl). http://www.steelnumber.com/en/steel_composition_eu.php?name_id=850, Last accessed 14 July 2022
13. Steel Navigator.: S355J2. https://steelnavigator.ovako.com/steel-grades/s355j2/, Last accessed 14 July 2022

Selection, Dimensioning and Maintenance of Roller Bearings

Eleonora Desnica[1(✉)], Aleksandar Ašonja[2], Ljiljana Radovanović[1], Ivan Palinkaš[1], and Imre Kiss[3]

[1] University of Novi Sad, Technical Faculty "Mihajlo Pupin", Zrenjanin, Republic of Serbia
desnica@tfzr.uns.ac.rs

[2] Faculty of Economics and Engineering Management in Novi Sad, University Business Academy in Novi Sad, Novi Sad, Republic of Serbia

[3] Faculty of Engineering Hunedoara, University Politehnica Timisoara, Timisoara, Romania

Abstract. Knowledge of issues related to the selection and dimensioning of roller bearings is extremely important in the work of engineers and designers, and from the point of view of the availability of available literature in this area, in general, is not at the level it should be. The main goal of this paper is to present the current procedures for the selection of roller bearings, to point out the most important factors to consider when choosing standard types of roller bearings, based on which their further selection can be significantly facilitated. Each type of bearing shows characteristic features, based on its design, which more or less make, or more or less correspond to the given conditions of application. Current methods and procedures for diagnosing the condition and maintenance of roller bearings as well as trends in their development will be presented. The paper deals with topics that show historical development of roller bearings, basic characteristics and types of roller bearings, selection and dimensioning of bearings, a diagnostic bearing condition inspection as a maintenance procedure is presented and described.

Keywords: Roller bearings · Selection and dimensioning of bearings · Condition diagnostic · Maintenance

1 Introduction—History of Roller Bearings

Bearings as elementary parts of every machine have a significant role in the functioning of machines, its maintenance and reliability, because they have the task of applying the load from the power transmission [1]. The reliability of the machine depends on the reliability of each of its individual parts, especially the moving components and elements, including roller bearings [2]. Bearings are standard assemblies and standards prescribed way of marking them [3]. The bearing designation consists of the basic markings that are fully stated in the technical documentation at the time of purchase and the supplementary markings that are listed as needed.

Little is known about the origin of this machine element, which is essential today for our daily social and professional life. Rolling as a physical phenomenon is almost as old

D. Blažević et al. (Eds.): OTO 2022, LNNS 592, pp. 133–142, 2023.
https://doi.org/10.1007/978-3-031-21429-5_12

as our civilization, and the first serious application is related to the period of construction of the first larger buildings, i.e. for the moment when a man decided to move those burdens that he himself could not carry. The first described mechanisms date from the 3rd century BC. At that time, various tools were known as simple machines, such as: wheel, pulley, winch, mill, screw, auger, gear, catapult, etc. Greek mathematicians and physicists of that time, such as Aristotle, Archimedes, Heron and others, are responsible for the analysis and realization of these devices [4].

The further development of the theory of mechanisms and machines went side by side with the development of society. The stagnation of the progress of scientific thought was evident from the beginning of the new era until almost the 14th century. There were no significant discoveries or great developments in science during that period. The improvement and development of all mechanisms and machines is closely related to the development of roller bearings. Today, almost no machine can be imagined without a roller bearing as the most vital work piece, starting with those light household machines all the way to heavy industrial machines.

We are witnessing from year to year that the same dimensions of certain types of roller bearings permanently reduce the weight of bearings, with constant improvements in terms of self-maintenance, higher load capacity, reduced vibration and noise, longer service life, etc. [5]. These and other similar reasons are very important from the aspect of further improvements in the process of designing roller bearings, therefore it is extremely important for designers to get acquainted with the development roller bearings throughout history [6] (Figs. 1 and 2).

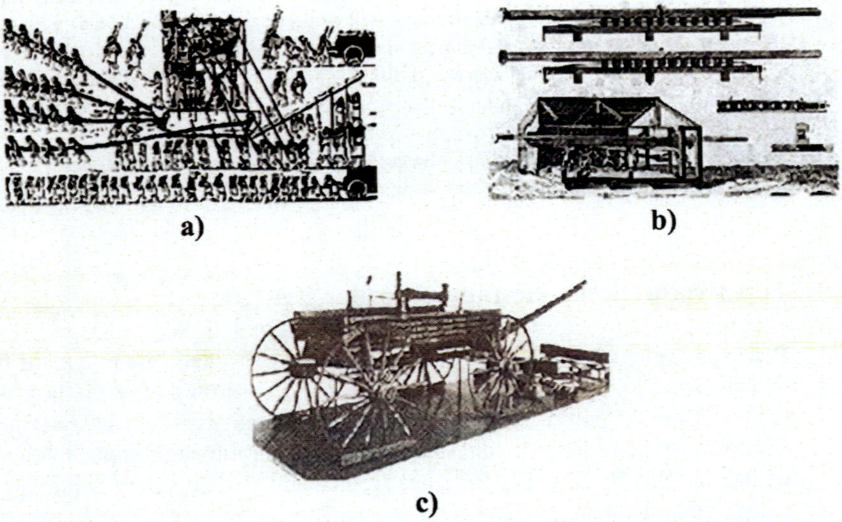

Fig. 1. Some of the earliest examples of rolling applications **a** the first example of the application of rolling (Assyria around 700 BC), **b** log for breaking (Greek design 330 BC) and **c** carriages similar in design to the "CART II" carriages (used between 300 and 100 BC) [7].

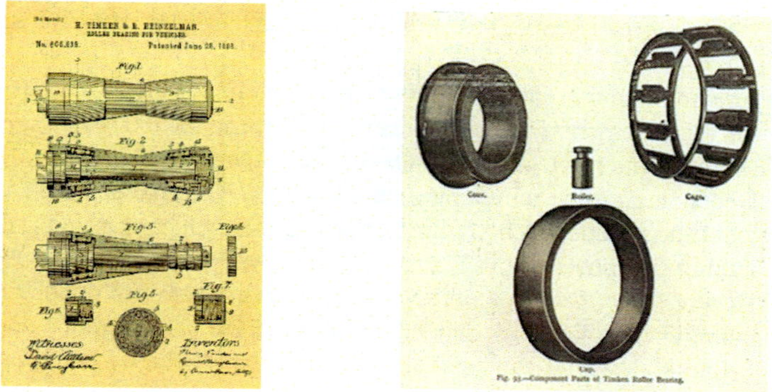

a) b)

Fig. 2. **a** Original drawing of the first patent of a tapered roller bearing, no. 606,635, from year 1898 [8] and **b** a drawing of a Timken tapered roller bearing (with cut rollers) [7].

2 Basic Characteristics and Types of Roller Bearings

Rolling bearings are used to provide conditions for the relative movement of rotating parts [9] and to transfer loads from moving to stationary parts or vice versa. In addition to sufficient bearing capacity, the bearing should provide the required accuracy of the position of the axis of the rotating parts in the state of rotation, high enough speed of rotation and to allow possible deviation of the position of the axis of rotation which may be due to deviations in production or elastic deformations. The extremely complex function of the bearings has imposed the need for the application of special manufacturing technologies and they are made by specialized manufacturers. A wide range of construction solutions has also been developed [10].

Shapes and dimensions of roller bearings, tolerances of dimensions and shapes, material, quality, method of installation and control are prescribed by standards. The bearings are delivered assembled and ready for installation, so that only the appropriate bearing is selected during construction (Fig. 3).

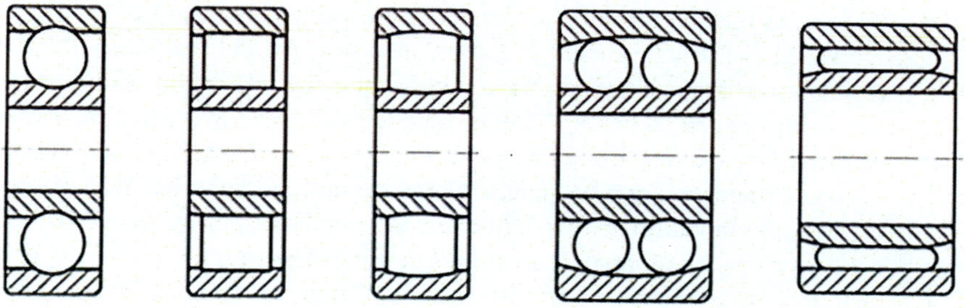

a) **b)** **c)** **d)** **e)**

Fig. 3. Rolling bearings: **a** ball bearing, **b** roller bearing, **c** barrel bearing, **d** self-adjusting double row ball bearing and **e** needle bearing [11].

The division of roller bearings can be done in several different ways:

- According to the direction of force, they are divided into: radial (transmitting force in the radial direction); axial (transmitting force in the axial direction) and radial-axial (transmitting force in the radial and axial directions) [12].
- Depending on the mounting method, they can be: fixed, which transmit radial and axial force in one or both directions, and free, which transmit only radial force and allow axial movement [13].
- According to the shape of the rolling elements, they are divided into: spherical and cylindrical (cylindrical, depending on the shape of the rolling elements, they can be cylindrical-cylindrical, conical-cylindrical, barrel-shaped and needle-shaped).
- According to the number of rows of rolling elements, the bearings can be: single-row and multi-row [7].
- Depending on whether they are disassembled or not (during installation), they can be: detachable and non-detachable.
- According to the ability to adapt to the deformation of the shaft in the support are divided into: rigid (non-adjustable) and articulated (adjustable).

3 Selection and Dimensioning of Bearings

The choice of bearing is conditioned by its application, i.e., operational characteristics and requirements. The choice of roller bearings can be made according to the load spectrum for a modified service life, where the probability of destruction can be chosen arbitrarily and where the operating conditions are taken into account.

The question is always present: What type of bearing to use? It is necessary to choose quality bearings adapted to the working conditions in which there will be no damage before the projected time of use, because only such a bearing will justify the funds invested in it and in the equipment.

When dimensioning bearings, it is necessary to set goals related to long service life, high safety and economy. In order to meet these goals, the designer needs to know the conditions in which the bearing works and the requirements it needs to meet. Environment, such as shaft, housing, fastening parts, sealing, lubrication, therefore, everything that makes the bearing suitable for the given application conditions [14].

When constructing shaft bearings, mechanisms and machines, the most relevant exploitation characteristic on the basis of which the selection of roller bearings is made is bearing capacity. One of the most important factors influencing the load distribution on the rolling elements, and thus on the bearing capacity, is the internal radial clearance.

The dimensioning flow should take place in a certain order. We start from the most precise determination of all in influential factors. The type, layout and size of the bearing are then chosen with many alternatives in mind. The following is a selection of specific bearings (with their markings, dimensions, tolerances, clearance, type of cage, etc.) for connection parts (fitting, fastening, sealing, etc.) and lubrication. When dimensioning the bearings, it is necessary to take into account the conditions of assembly and disassembly of the bearings.

When dimensioning bearings, it is primarily necessary to know the influences [7]:

- Machinery, mechanisms or place of bearings,
- Working conditions (forces, speed of rotation, installation space, temperature, conditions environment, shaft position, stiffness of elements, etc.),
- Various requirements (service life, accuracy, noise, friction, operating temperature, lubrication, maintenance, assembly and disassembly, etc.) and
- Commercial benefits (cost limits, economic justification, quantities, etc.).

All of the impacts need to be familiarized in detail before approaching bearing sizing. Sizing of bearings consists of defining the following phases:

- Choice of bearing type,
- Choice of bed layout,
- Determination of bearing size (service life and static safety),
- Determination of other bearing parameters,
- Determination of other parts in assembly,
- Choice of lubrication and maintenance methods and
- Determining the method of assembly and disassembly.

4 Lubrication and Maintenance of Bearings

In addition to the inner ring, outer ring, roller elements and cage, each roller bearing also has a fifth element—lubricant. Unfortunately, in practice, the lubricant is not given much attention. This is best shown by the statistics that almost 50% of all damage to the bearings comes from inadequate lubrication.

The service life of roller bearings is significantly reduced if the thickness of the oil film is not sufficient to protect the metal-to-metal contact [15]. One of the ways to overcome this is certainly the selection and use of appropriate lubricants or additives that eliminate high temperatures caused by local contacts and sealants that prevent the entry of impurities. The use of these additives means an increase in the wear of the oil film at the points of contact, the end result of which is lower contact pressure and an increase in the lifespan of the bearings [13].

Long-term practice has shown that every third roller bearing fails due to poor lubrication, which certainly raises the question of justification for the use and improvement of new technologies in production and procedures in calculating tribological parameters for roller bearings. From the aspect of improving the lubrication of roller bearings, the emphasis is increasingly placed on constant increases in the performance of materials and lubricants [16].

The general opinion is that the performance of today's bearings is significantly higher than those obtained on the basis of the nominal age calculation. The reason for this should be sought in the current trends that are present in the development of bearings. Bearing development activities can be related to:

- Research and improvements in the field of materials,
- Improvement of construction and production technology,
- Clear definition of bearing operating parameters,

- Environmental impact analysis and
- **Bearing sealing efficiency.**

Improvements made in the field of roller bearings have reflected their increasing reliability and longer service life. However, further progress in the development of roller bearings is limited by tribological barriers, so current research focuses on solving tribological problems, primarily on the application of EHD lubrication theory (macro and micro) **clearer definitions of surface fatigue causes, and analysis of the impact of working** conditions and environment on roller bearing damage.

The following lubricants are used to lubricate roller bearings:

- Fats (oils + thickeners) in 90%,
- Oils in 9% and
- **Other lubricants up to 1% (graffiti MoS_2**, sliding varnishes, silver, etc.) (Fig. 4).

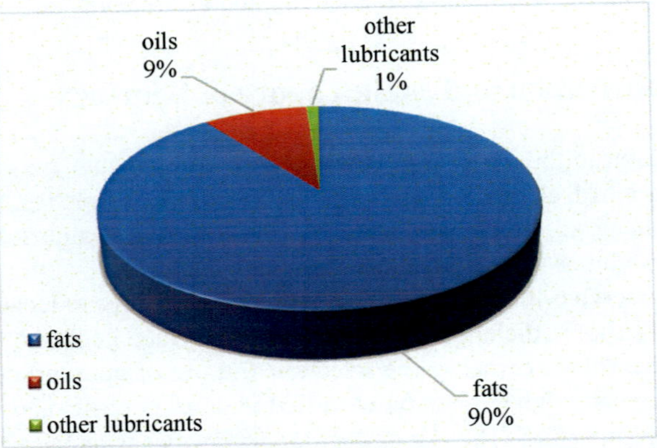

Fig. 4. Percentage display of used lubricants for lubrication of roller bearings [7].

The application of modern methods and procedures in the maintenance of roller bearings, multifunctional solutions for sealing and lubrication, has certainly paved the way for safer and more efficient operation of bearings as prevailing in industrial production. Further operation and maintenance of bearings should primarily indicate the reduction of damage to roller bearings by introducing new seals and lubricants in operation or increasing their lifespan.

Based on the experience gained so far, it can be freely concluded that the service life of roller bearings is unpredictable. The possibility of failure of this element is very high. Analyzes show that possible deviations range up to 20 times in relation to the theoretical service life. However, with the right choice of bearings, adequate operation and maintenance, they can only move in a positive direction.

5 Diagnostic Inspection of Bearing Condition

The diagnostic inspection of the bearing condition will be shown on the example of monitoring the operating conditions of the annular ball bearing 6212 on the centrifugal pumps using the software program "Lubmaster". Shock analyzer pulse A2011 every three months the condition of the bearing was monitored. Measurement results, such as bearing temperature and other required lubrication data entered were used for evaluation modified nominal bearing life. In the program "Lubmaster" for estimating the lifetime using the "Life time graph", based on the calculated value of the kappa factor (κ), the factor *a23* was shown for materials and lubricants, which was necessary for calculation modified nominal bearing life in hours. In Figs. 5, 6 and 7 is given representation of the modified nominative age *L10ah* and other sizes. In the first measurement for bearing temperature of 82 °C, the modified nominal age was 24.000 h, in the second measurement for a bearing temperature of 84 °C, the modified nominal age was 22.000 h and in the third measurement for the bearing temperature of 92 °C, the modified nominal age was 16.000h.

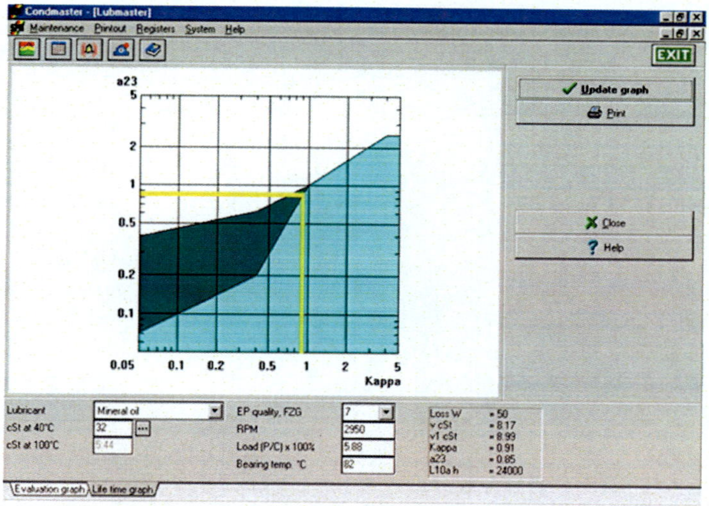

Fig. 5. Modified nominal life values on bearing 6212—first measurement.

The input values that are entered into the software when estimating the service life of the bearing are:

- Lubricant—the type of lubricant is selected from the menu: mineral oil, synthetic oil and polyglycol,
- cSt at 40 °C—viscosity of lubricant in cent stocks at 40 °C,
- cSt at 100 °C—viscosity of lubricant in cent stocks at 100 °C,
- EP qual FZG—(FZG) numbers express the quality of the high-pressure additive in the lubricant, (for oils without additives the FZG = 7) and
- Load ratio P/S × 100—percentage amount of load capacity.

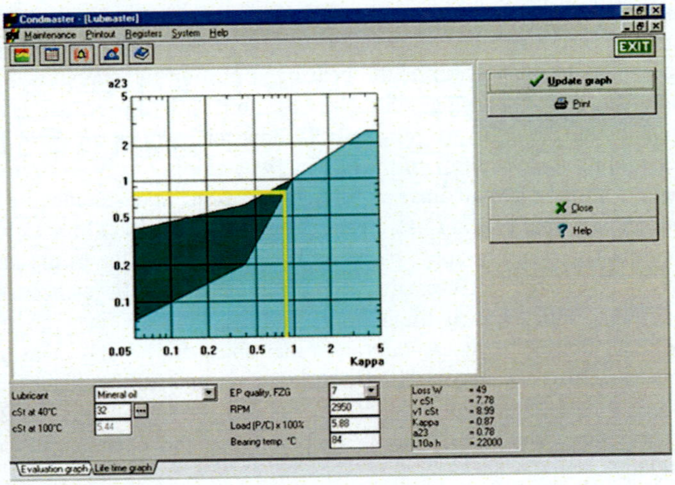

Fig. 6. Modified nominal life values on bearing 6212—second measurement.

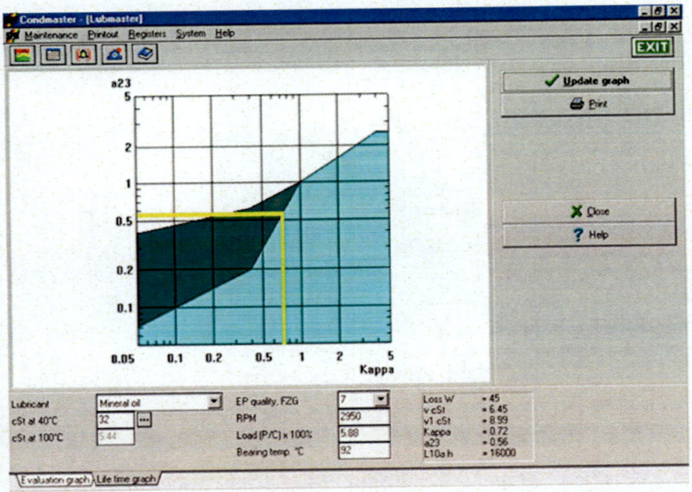

Fig. 7. Modified nominal life values on bearing 6212—third measurement.

The output values obtained after data processing were:

- Bearing temp °C—bearing temperature, ranging from (0–200) °C,
- Loos—friction loss, in relation to bearing type, lubricant load and viscosity (W),
- ν—working viscosity of the lubricant (kinematic viscosity), depends on the bearing temperature, can be obtained from the tables of the lubricant manufacturer (cSt),
- ν—viscosity required to create an oil film, allowing prevention of excessive wear (cSt),

- kappa—a factor that describes the oil film (-), in the literature is often denotes as a factor κ,
- a_{23} durability factor (-) and
- L_{10ah}—modified nominal bearing life (h), (max. 900.000 h).

6 Conclusion

Rolling bearings are unique machine elements, which have an almost paradoxical combination of unity and inexhaustible possibilities for further improvements. Seemingly simple construction, which consists of only four elements, roller bearings in the micro domain open a huge space that extends through the entire engineering disciplines. It is on this fact that large bearing manufacturers build their technological excellence.

The process of developing new technologies in the field of design and production of roller bearings today is a very inspiring research task. This is most pronounced primarily in design, where good financial effects and great time savings can be achieved. We are witnesses that for the same dimensions of certain types of roller bearings from year to year, the mass of bearings is permanently reduced, with constant improvements in terms of self-maintenance, higher load capacity, reduced vibration and noise, longer service life, etc.

The application of any method for diagnostic analysis of bearing condition, indicates better information about the condition of roller bearings and other parts on technical systems, which allows us to raise much greater availability and functionality of technical systems. New methods and techniques in the field of technical diagnostics allow us to better plan, reduce overtime, significantly increase the time between two failures, and knowing the condition of bearings on technical systems allows us to increase profits.

References

1. Mehmet, I., Dogan, G., Chitariu, D.F., Dumitraş, C., Negoescu, F.: Research on advances in roller bearing manufacturing. IOP Conf. Ser. Mater. Sci. Eng. **1182**, 012045 (2021). https://doi.org/10.1088/1757-899X/1182/1/012045
2. Janjic, N., Adamovic, Z., Nikolic, D., Asonja, A., Stojanovic, B.: Impact of diagnostics state model to the reliability of motor vehicles. J. Balkan Tribological Assoc. **21**(2), 511–522 (2015)
3. Mikic, D., Desnica, E., Asonja, A., Stojanovic, B., Epifanic-Pajic, V.: Reliability analysis of ball bearing on the crankshaft of piston compressors. J. Balkan Tribological Assoc. **22**(4-IV), 5060–5070 (2016)
4. Gligorić, R.: Mechanisms of Agricultural Machinery. University of Novi Sad, Faculty of Agriculture, Novi Sad, (2005)
5. Gülbahçe, E., Çelik, M.: Investigation of piezoelectric patches placement according to the strain intensity. Appl. Eng. Lett. **5**(2), 46–49, (2020). https://doi.org/10.18485/aeletters.2020.5.2.2
6. Ašonja, A., Adamović, Ž., Gligorić, R.: Development of roller bearings throughout history. In: XXXIII May Meeting of Maintainers "Teleautomation of Machines and Plants in Industry—Informatics and Ecology". Vrnjačka Banja (2010)

7. Ašonja, A., Adamović, Ž., Gligorić, R.: Selection and Dimensioning of Roller Bearings. Duga knjiga, Sremski Karlovci (2010)
8. Timken Roller Bearing Patent Art 1898, No606,635, Art Print, by Ian Monk
9. Zarei, A., Farahani, S., Pradeep, S.A., Driscoll, J. Lukasiewicz R., Pilla S.: Design and manufacturing of roller bearing polymeric cages and development of a theoretical model for predicting the roller push-out force. Sci. Rep. **12**, 1017 (2022). https://doi.org/10.1038/s41598-022-04959-9
10. Patil, A.A., Desai, S.S., Patil, L.N., Patil, S.A.: Adopting artificial neural network for wear investigation of ball bearing materials under pure sliding condition. Appl. Eng. Lett. **7**(2), 81–88 (2022). https://doi.org/10.18485/aeletters.2022.7.2.5
11. Gligorić, R.: Machine elements. University of Novi Sad, Faculty of Agriculture, Novi Sad (2015)
12. Mikić, D., Desnica, E., Kiss, I., Mikić, V.: Reliability analysis of rolling ball bearings considering the bearing radial clearance and operating temperature. Adv. Eng. Lett. **1**(1), 16–22 (2022). https://doi.org/10.46793/adeletters.2022.1.1.3
13. Ašonja, A., Adamović, Ž.: Maintenance of Roller Bearings. Duga knjiga, Sremski Karlovci (2010)
14. FKL catalog, Feljton, Novi Sad (2002)
15. Lundberg, J., Parida, A., Söderholm, P.: Running temperature and mechanical stability of grease as maintenance parameters of railway bearings. Int. J. Autom. Comput. **7**(2), 160–166 (2010). https://doi.org/10.1007/s11633-010-0160-1
16. Adamović, Ž, Adamov, J.: Tribology and Lubrication. Society for Technical Diagnostics of Serbia, Smederevo (2006)

Methode for Quick Determination of the Reliability Level of Agricultural PTO Shafts

Aleksandar Ašonja[1], Eleonora Desnica[2(✉)], Alexander Pastukhov[3], Yury Kuznetsov[4], and Igor Kravchenko[5,6]

[1] Faculty of Economics and Engineering Management in Novi Sad, University Business Academy in Novi Sad, Novi Sad, Republic of Serbia
[2] University of Novi Sad, Technical Faculty "Mihajlo Pupin", Zrenjanin, Republic of Serbia
desnica@tfzr.uns.ac.rs
[3] Belgorod State Agricultural University named after V. Gorin, Mayskiy, Russian Federation
[4] Orel State Agrarian University named after N.V. Parakhin, Orel, Russian Federation
[5] Institute of Mechanical Engineering of the Russian Academy of Sciences named after A.A. Blagonravov (IMASH RAS), Moscow, Russian Federation
[6] Russian State Agrarian University—MTAA named after K.A. Timiryazev, Moscow, Russian Federation

Abstract. The aim of the presented research was to test and develop a new method for rapid determination of the reliability of agricultural PTO shafts, based on the diagnosis of the condition of bearing assemblies on agricultural PTO shafts. The paper presents a method based on the rapid determination of the reliability of agricultural PTO shafts used in operation. The method is based on testing the temperature in the PTO shaft bearing assemblies. The main features of this method are the simple operation of the testing device and measuring equipment, very short testing time lasting 4.02 min/sample, low energy consumption for tests of 1.005 kW/sample and mobility of the testing device.

Keywords: Method · Reliability · Temperature · Laboratory device · PTO shaft · Bearing assembly

1 Introduction

In relation to other industries, agricultural production is a specific activity. In it, agricultural machines are used to work in unfavorable working conditions, and they are engaged mainly seasonally [1]. In that period, they are very intensively used in agro-technical processes: sowing, harvesting, reaping, etc. The failure of any vital part of an agricultural machine (e.g., roller bearing failure) is the failure of the whole machine [2]. And if all that happens e.g., during the harvest, the damages of downtime are very great and can far exceed the price of the roller bearing, PTO shaft, and the whole machine. Therefore, in recent years, in agricultural technology more and more attention has been paid to product design [3], material selection, welding method, [4], etc.

D. Blažević et al. (Eds.): OTO 2022, LNNS 592, pp. 143–149, 2023.
https://doi.org/10.1007/978-3-031-21429-5_13

Conditions for the agricultural PTO shafts' use are extremely unfavorable, both due to the permanent load and speed increase on agricultural machinery [1] and due to their inadequate practical use [5]. During exploitation, they experience a state of high dynamic load (impact) [6], which certainly adversely affects the bearing assemblies in the universal/Cardan joint. Researches on problems in this area are intensive in the world, due to the need of increasing the reliability of agricultural machinery, in order to avoid the cancellation of their work during a seasonal job that needs to be done. Tribological processes of friction and tear and other damages to roller bearings, caused primarily by the occurrence of vibration and temperature's high levels in bearing assemblies during operation are inevitable, which should be foreseen and eliminated in time, by applying technical maintenance [7–9], repair and overhaul [10–13].

The reliability of the tractor PTO shaft directly affects the quality and efficiency of agricultural processes [14]. Increasing the reliability of PTO shafts in operation could lead to improving their quality of work, saving resources and time [1, 2], improvement of safety and ergonomic working conditions for operators [15–18], etc. The aim of the research in this paper was to develop a new method for fast determination of the reliability of agricultural PTO shafts, based on the diagnostics of the condition of bearing assemblies on PTO shafts.

2 Material and Method of Work

The method for rapid testing of the current reliability of agricultural PTO shafts is based on testing the temperature in the bearing assemblies of the Cardan joints of agricultural PTO shafts Fig. 1. The method is suitable/recommended for determining the current reliability of exploitation samples of agricultural PTO shafts [19, 20]. Laboratory and measuring equipment necessary for testing:

- Laboratory testing device: laboratory test table (Model: ANA, Type: 23-26-26-04) with electric brake (Model: EC, Type: 3/28),
- Temperature measuring equipment: Laser infrared thermometer,
- Computer timekeeping and data processing: software for measuring data adapted to the examination of two joints.

To test the method, 10 samples of agricultural PTO shafts of different age periods were used. The technical and technological characteristics in which the method was tested were:

- Size I of double agricultural PTO shafts was tested,
- PTO shafts of different age periods brought out of operation were tested,
- The angle of rotation of the cardan joints was $\alpha_{12} = \alpha_{21} = 20°$,
- The PTO shafts "Z" were tested horizontally,
- The PTO speed was 540 min^{-1},
- The load on the brake was ~75 Nm.

A method for rapid testing of the current reliability of agricultural PTO shafts has been developed and applied to specially developed laboratory equipment on a laboratory

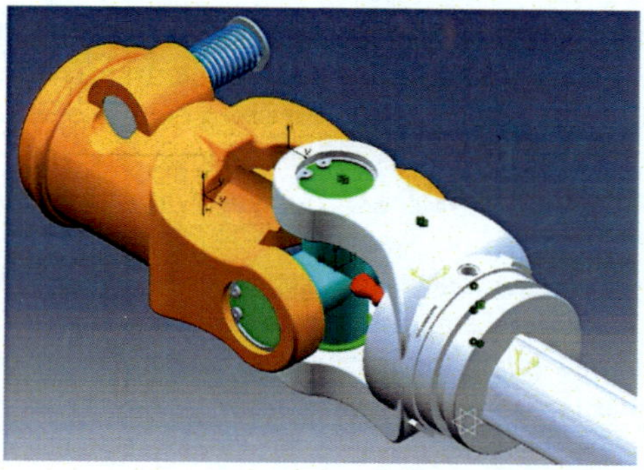

Fig. 1. Tests of cardan joint bearings. (*Source* Authors)

table (Model: ANA, Type: 23-26-26-04) on which there is an electric brake (Model: EK, Type: 3/28) which is used for simulation of load during testing, Fig. 2. The laboratory table is intended for testing the reliability of agricultural PTO shafts, although it can also be used for testing other PTO shafts and other mechanical power transmissions. The laboratory table itself is presented in detail in the research [1, 2, 21]. Examples of laboratory tables with an intermediate part are given in the research [22].

On the presented laboratory equipment, measurements of temperature parameters in PTO shaft joints can be: manual, automated, or combined. Sensors can be used for fully automated tests, and research results can be continuously collected and stored in one. However, since the main feature of the presented method is the speed of testing, it is most convenient to perform it by manual non-contact measurement with a laser infrared thermometer, as was the case when testing the method.

Fig. 2. Testing the current reliability of agricultural PTO shafts [19].

If the test time is not a limiting factor, it can be done as already mentioned with measuring sensors that would be placed on the bearing units, where the temperature from the bearing assemblies would be directly transferred to the computer. Experience has shown that these tests take much longer to transfer the temperature from the joint to the bearing units, where there is always a danger of differences in the values of the temperature of the bearing units and the bearing assemblies of the Cardan joints.

Experimental final variables (set) values when testing the current reliability of agricultural PTO shafts can be: loads at the beginning of the PTO shaft (depends on the revolutions' number and electric motor's power), loads/torques at the end of the shaft (depends on shaft speed and electric brake power) and angle of rotation.

The measuring points on the bearing assemblies of the PTO shafts on which the temperature was measured are shown in Fig. 3. (pos. 5 and 7). The arrangement of other mechanical power transmissions on the laboratory table is shown in Fig. 4. (1—input pulley on the drive, 2,4,8,10—fixed bearing units, 3—the first auxiliary shaft, 5—the first joint, 6—tested PTO shaft, 7—the second joint, 9—the second auxiliary shaft, 11—brake output pulley).

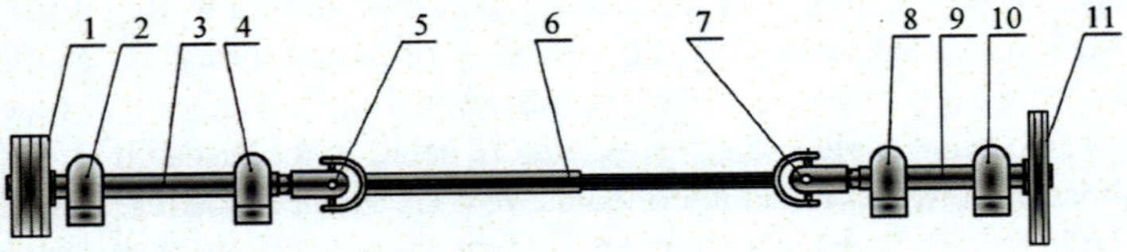

Fig. 3. Arrangement of all mechanical power transmissions on the laboratory table [19].

The examination of the current reliability of agricultural PTO shafts is based on the diagnostics of bearing assemblies temperature on Cardan shaft joints, which depends on the amount of damage to bearing elements (needle bearings, crosses, cups, etc.). As a consequence of these damages, radial, axial and circular clearances appear, which lead to nonlinear contact (inadequate lubrication) in the bearing assemblies, thus leading to an increase in temperature in them. During the test, as a final assessment of the reduced reliability of the Cardan joints (damage to the bearing assemblies on the Cardan joints) and the interruption of the test, the moment was taken when:

- The temperature in the bearing assemblies (at least one joint) exceeded 73 °C—which is characterized by the state of failure of the PTO shaft and
- The temperature in the bearing assemblies did not increase by more than 10 °C for 3 min of testing—which is characterized by the correct condition of the PTO shaft.

3 Research Results

The approximate test time of the current reliability ranged from 3.41 to 4.48 min/sample, Fig. 4. By testing a method for rapid testing, which is based on the technical diagnostics of the PTO shaft bearings, it was concluded that 60% of the tested shafts are defective, Fig. 5. Conducted tests on random samples of agricultural PTO shafts showed a very low operational reliability.

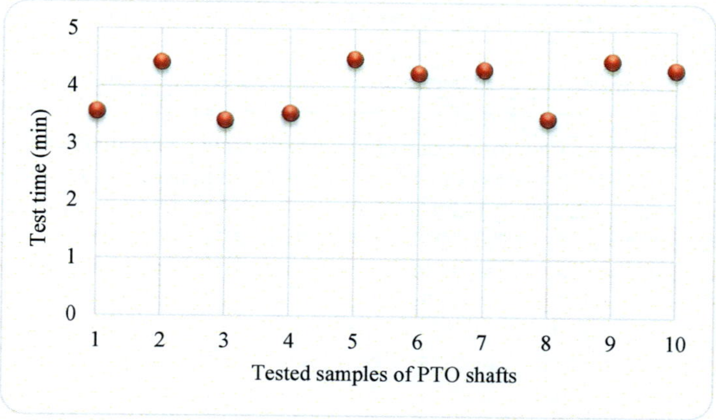

Fig. 4. Average time of diagnostics of tested samples of PTO shafts.

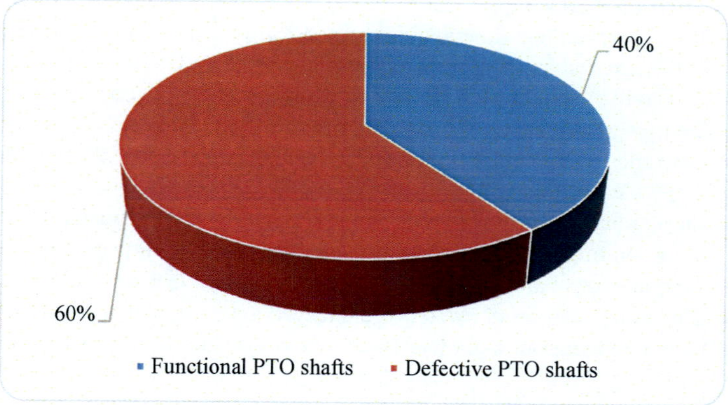

Fig. 5. Percentage testing of agricultural PTO shafts.

Based on the conducted research, a decision can be made to withdraw the damaged PTO shaft from the operation or to repair it, which directly affects the increase of safety of all operators of agricultural equipment and the reliability of other agricultural attachments powered by PTO shafts.

4 Conclusion

By applying the given technical solution, ie by applying the experimental table for testing the reliability of agricultural PTO shafts, it is possible to monitor the behavior of all types of PTO shafts (single or double with Z or with W working performance) during operation. The presented technical and technological characteristics under which the test was performed are identical to the conditions in which the PTO shafts are operated in the field: the angle of rotation of the output and input shafts, the loads on the output shaft, the number of turns on the input shaft.

Clogged channels for lubrication and damage (breakage) of needle bearings stand out as the main causes of this condition of bearing assemblies on PTO shafts. Very low reliability is a consequence of inadequate technical maintenance and use, as well as complex conditions prevailing in agricultural production. However, training operators of agricultural machinery in terms of use and maintenance would increase the operational level of reliability of PTO shafts and could significantly increase their safety.

The main advantages of the method for rapid testing of current reliability are:

- The current reliability of exploitation samples of agricultural PTO shafts is being tested for the first time,
- Very simple test method,
- The universality of technical-technological testing criteria,
- Short reliability test time (from the moment of installation of the tested samples, starting of the drive and brake, to the moment of testing) of only 4.02 min/sample,
- Up to 12 shaft samples can be tested in one hour, ie. For 8 h 96 samples,
- The shortened test time avoids the possibility of external influences on the accuracy of the test results (heat transfer from the brake and drive part),
- Due to the very short test period, it is not necessary to cool the brake,
- Savings in human labor and energy required for testing,
- Low energy consumption required for the test of 1.005 kW/sample, ie for the drive 0.737 kW/sample and for the brake 0.268 kW/sample),
- The application of the method in practice can significantly affect the safety of operators of agricultural equipment and the reliability of other attachments in agriculture,
- Simple, cheap, and easily accessible temperature testing device,
- The entire laboratory system is characterized by device mobility and easy switching from one location to another (from one diagnostic center to another).

The advantage of the presented method lies in the fact that by adapting the connection elements on the laboratory table with appropriate technical and technological characteristics of testing, it is possible to perform tests of other types of PTO shafts, with limiting transmission dimensions and meeting the load range.

References

1. Ašonja, A., Adamović, Ž., Jevtić, N.: Analysis of reliability of cardan shafts based on condition diagnostics of bearing assembly in cardan joints. Metal. Int. **18** (Spec.4), 216–221 (2013)
2. Ašonja, A.N., Pastuhov, A.G.: The influence of diagnostic state of reliability of agriculture double cardan shaft. Agric. Eng. **39**(1), 21–30 (2014)
3. Upadhyay, G., Raheman, H., Rasool, S.: Three-dimensional modelling and stress analysis of a powered single acting disc harrow using FEA. Curr. Agric. Res. 5(2) 203–219 (2017). http://dx.doi.org/10.12944/CARJ.5.2.08
4. Kuznetsov, M.A., Il'yashchenko, D.P., Lavrova, E.V., Solodsky, S.A., Podgornyh, O.A., Verkhoturova, E.V.: Determination of optimal modes for the production of micro range metal powders. Appl. Eng. Lett. **7**(1), 17–24 (2022). https://doi.org/10.18485/aeletters.2022.7.1.3

5. Corte Bellochio, S.D., Santos Alonço, A. dos., Possebom, G., Gubiani, P., Pagliarin, L.: Presence and dimensional assessment of power take-off (PTO) protection of new Brazilian agricultural tractors. CIGR J. **21**(3), 130–137 (2019)
6. Celik, H.K., Cinar, R., Kunt, G., Rennie, A.E.W., Ucar, M., Akinci, I.: Finite element analysis of a PTO shaft used in an agricultural tractor. Ergonomics Int. J. **2**(3), 000147 (2018). https://doi.org/10.23880/eoij-16000147
7. Novaković, B., Desnica, E., Radovanović, LJ., Ivetić, R., Đorđević, L., Labović Vukić, D.: Optimization of industrial fan system using methods laser alignment. Appl. Eng. Lett. **6**(2), 62–68 (2021). https://doi.org/10.18485/aeletters.2021.6.2.3
8. Mikic, D., Desnica, E., Asonja, A., Stojanovic, B., Epifanic-Pajic, V.: Reliability analysis of ball bearing on the crankshaft of piston compressors. J. Balkan Tribological Assoc. **22**(4-IV), 5060–5070 (2016)
9. Desnica, E., Ašonja, A.: Damage of rolling bearings on agricultural machines. J. Tractors Power Mach. **23**(3–4), 33–38 (2018)
10. Pastukhov, A., Timashov, E., Kravchenko, I., Parnikova, T.: Thermometric diagnostics of elements of reaper drive for sunflower harvesting. In: 21st International Scientific Conference "Engineering for Rural Development", pp. 43–48. Jelgava, Latvia (2021). https://doi.org/10.22616/ERDev.2021.20.TF008
11. Janjic, N., Adamovic, Z., Nikolic, D., Asonja, A., Stojanovic, B.: Impact of diagnostics state model to the reliability of motor vehicles. J. Balkan Tribological Assoc. **21**(2), 511–522 (2015)
12. Ašonja, A.: Economic justification of bearing assembly repair. Agric. Tech. **33**(1), 67–73 (2008)
13. Desnica, E., Jakovljević, V., Kiss, I., Radovanović, L., Đorđević, L.: Techno-economic analysis of the justification of the machine part reparation procedure on the excavator. In: Glavaš, H., Hadzima-Nyarko, M., Karakašić, M., Ademović, N., Avdaković, S. (eds.) 30th International Conference on Organization and Technology of Maintenance (OTO 2021). OTO 2021. LNNS, vol. 369, pp. 92–103, Springer, Cham, Switzerland (2022). https://doi.org/10.1007/978-3-030-92851-3_7
14. Wang, Y., Wang, L., Zong, J., Lv, D., Wang, S.: Research on loading method of tractor PTO based on dynamic load spectrum. Agriculture **11**, 982 (2021). https://doi.org/10.3390/agriculture11100982
15. Ašonja, A., Desnica, E.: Research into reliability of agriculture universal joint shafts based on temperature measuring in universal joint bearing assemblies. Span. J. Agric. Res. **13**(1), 1–8 (2015). https://doi.org/10.5424/sjar/2015131-6371
16. Varani, M., Mattetti, M., Molari, G.: Performance evaluation of electrically driven agricultural implements powered by an external generator. Agronomy **11**, 1447 (2021). https://doi.org/10.3390/agronomy11081447
17. Ašonja, A., Cvetković, S., Mikić, D.: Testing of cardan shaft operating reliability in agricultural machines. Tech. Diagn. **12**(3), 12–19 (2013)
18. Ašonja, A., Adamović, Ž.: Maintenance of rolling bearings. Duga knjiga, Sremski Karlovci, Serbia (2010)
19. Ašonja, A., Gligorić, R. Pastukhov, A.: The method for rapid testing of current reliability of agricultural cardan shafts. Technical Solution. Novi Sad, Serbia (2014)
20. Ašonja, A., Adamović, Ž. Mikić D.: The method for the rapid determination of the confidence level of agricultural cardan shafts. Tech. Diagn. **6**(1–2), 30–36 (2014)
21. Desnica, E., Ašonja, A., Mikić, D., Stojanović, B.: Reliability of model of bearing assembly on an agricultural of Cardan shaft. J. Balkan Tribological Assoc. **21**(1), 38–48 (2015)
22. Pastukhov, A., Timashov, E., Sharaya, O., Bakharev, D.: CAE—Justification of the leading shaft of the test stand. In: 7th TAE 2019, pp. 429–434. Prague, Czech Republic (2019)

Regeneration of Mould Inserts Using the Laser Welding Method

Waldemar Matysiak[1], Miroslav Duspara[2](✉), Grzegorz Stawiński[3], Josip Cumin[2], Karol Malutkiewicz[3], Antun Stoić[2], and Jakub Hajkowski[1]

[1] Poznan University of Technology, Poznan, Poland
[2] University of Slavonski Brod, Slavonski Brod, Croatia
mduspara@unisb.hr
[3] GTS-GS, Lipno, Poland

Abstract. This paper concerns the wear of the working area of the permanent mould is strictly related to downfall in its functional properties. It may be caused by several different physical and chemical processes occurring during its exploitation and resulting mainly from the working conditions. The simplest wear factor is the material loss in the functional layer of the tooling, resulting mainly from the plastic strain, friction, corrosion or cracks of the surface layer. The tests were focused on comparative tests of inserts for high-pressure die-casting molds after their regeneration. Two methods were used for this process: TIG welding and laser surfacing. The scope of research included the study of surface microtopography and material structure on metallographic specimens. In particular, the thickness of the overlay layer, the heat affected zone were examined.

Keywords: Regeneration of mould inserts · TIG welding · Laser welding

1 Introduction

Wear of machine and device elements represents the reduction in properties of working area qualities. It can be caused by multiple physical and chemical processes that occur during the product lifetime. Wear mechanisms are very complex, thus containing many correlated factors. Influence intensity of above-mentioned factors depends on type of working environment, type and values of work parameters of the machine parts [1, 2].

Designers and manufacturers of machines and devices are striving to assure possibly best technical and economic results, continuously increasing the working parameters. However, the development in the material engineering, metallurgy and welding technology provides better and better metal, ceramic, cermetal and plastic materials, which easily conform to still increasing requirements, put on devices and additionally increase their durability. The problem of machine wear and tear has to be taken into consideration in consequent stages of design, manufacturing and exploitation. Designer in its assumptions has to apply different safety factors and construction durability and together with technologist have to specify type of material that meets these requirements.

D. Blažević et al. (Eds.): OTO 2022, LNNS 592, pp. 150–159, 2023.
https://doi.org/10.1007/978-3-031-21429-5_14

Manufacturer, simultaneously with equipment production process, conducts the production of spare parts that is used by the final equipment user [3, 4]. In the coming years and in accordance with industrial rule "zero faults", environment protection and heading for maximum technical and economic outputs, engineer will have to assure maximum machine and device durability, incurring possibly lowest cost [1, 5]. Welding engineers have special job to do as welding techniques assure lowest production and exploitation cost.

Practically each used part may be put through regeneration process using one of many available welding techniques. Planned regeneration is even a better solution—design of machine and device parts, which functional surfaces are manufactured using materials of special properties with possibility of adding these materials multiple times when certain level of wear and tear is reached.

Development in material engineering, welding technology, electronics and electrotechnics enables an overlaying welding of parts manufactured out of all known construction materials, including glass and wood, of any shape—from micro circuits to begin with, to heavy excavators. It can be carried out manually, half-automatically, automatically or in a robotized way, under roof and in the field. Knowledge of properties and scope of application of modern constructions materials can be used only under condition that processes of overlaying welding are also well known at the same time.

The article describes the causes of wear of mould parts, the properties of metal welding materials and welding techniques used for adding layers of special functional qualities.

2 State of the Art. The Reasons for Wear and Tear of Dies

The wear of the working area of the permanent mould is strictly related to downfall in its functional properties. It may be caused by several different physical and chemical processes occurring during its exploitation and resulting mainly from the working conditions. Wear mechanisms are very complex and cover many self-related factors, which of most importance are [1, 6, 7]: type and size of mechanical load, slide velocity between surfaces where friction occurs (surface layers), permanent mould work temperature, hardness and structure of friction surfaces, roughness factor of working surfaces, corrosive environment, friction factor of the working surfaces, wear process period. In the articles [8–10] the authors investigated how machined surface roughness affects friction between contact pairs. Friction in newly machined tool is lesser than in the tool which has a number of working hours on it (due to the abrasion and heat cracks).

The authors [11] have shown how laser welding process parameters change surface roughness on aluminium alloys which is comparable to the cladding procedure shown here. The paper [12] has similar research, but on Ti-6Al-4V material.

Working surfaces of tooling are often prior to wear due to separate or combined influence of: friction processes—abrasive wear and adhesive wear (tribological), shock load, high temperatures, erosion and cavitation, corrosion. The simplest wear factor is the material loss in the functional layer of the tooling, resulting mainly from the plastic strain, friction, corrosion or cracks of the surface layer [1, 7, 8].

The analysis of the reasons for wear of the machine and devices shows that ca. 50% of parts are subject to abrasive wear, 15% are subject to adhesive wear, 8%—erosion, 8%—are subject to fretting, 5%—subject to wear due to corrosion and approx. 14%—combined influence of e.g. corrosion, erosion and abrasion [8].

In order to make a rational choice of additional materials and establishing the technical conditions of adding layers that assure maximum durability of welded components with minimum cost incurred it is necessary to understand each phenomenon that accompany each of these wear processes. Machine and device components that are made out of metal or plastics even though lack of any exploitation load, stored in warehouse, in the vacuum-tight package that protects from atmospheric corrosion and UV radiation are due to natural ageing. Depending on the temperature of steel component, in the period of few months up to few years of storage a loss of or decrease in initial physical qualities may occur. It is caused by collapse of oversaturated solid solution, closing the metal composition to the state of equilibrium, that leads to decrease in plasticity in case of carbon steel. When it comes to permanent moulds, the adhesive wear is the one that occurs most of the time through grafting of first type, which happens when friction of metal surfaces without lubricants is the case [3].

Metal material that are based on iron matrix, most often used as material for machine and device parts are showing high inclination towards grafting and adhesive merge, as well as towards creation of oxide layers that protect from adhesive wear. Therefore intensity of adhesive wear of mating components depends on friction conditions that influence the rate of creating oxide layers. Typical machine parts that are prior to adhesive wear are metallurgic roller, cutting blades, shafts, pins and all kind of components that have bearing surfaces which are not lubricated. This type of friction is referred to as metal-to-metal friction type. The smallest wear is assured by low-alloyed, hard martensitic steel and manganese austenitic steel, as well as alloys based on cobalt and nickel matrix. The right choice of both friction metals is very important. If one of the metal has much harder than the other one, then the one which is less harder will be worn out much faster. Beside the described basic processes of tribological wear, in specific exploitation conditions of permanent moulds, there may be following types of wear [4]: oxidation wear, fatigue wear, spalling, high temperature wear, erosion wear.

Oxidation wear; occur during sliding friction and rolling friction. It shows as destruction of surface layer during the friction due to separation of oxide layers created through absorption of oxygen from atmosphere in the area where friction occurs. Oxygen diffusion follows in the micro-volumes of metal which is subject to elastic and plastic strain, with simultaneous creation of solid solution layers. Wear process requires that intensity of creating oxide layers was greater than intensity of surface degradation by friction process.

Fatigue wear occurs as a result of cumulating of deformations in the surface layer, caused by combined reaction of internal and external stresses. As an effect in the areas affected by stress pile-up, fatigue microcracks are being created, that evolve during the exploitation into macrocracks. Loss of coherence follows and metal molecules are torn out after reaching the border of fatigue strength by each metal microvolume of surface layer and related component weight loss.

Spalling occurs as gradual increase in stresses within surface layer of two components that are due to friction without lubrication, as a result of periodical influence of contact stresses and consequently leading to fatigue microcracks. Further periodical increase of stresses in the borders of Hertz stresses leads to spread of microcracks and falling from surface layer of friction components.

High temperature wear is process of constant decrease in functional properties not only when it comes to work surface but also regarding whole component. Prolonged influence of high temperatures depending on work conditions causes accelerated high-temperature corrosion and irreversible structural changes in the whole heat-affected component or at the depth of heat-affected zone (HAZ). As an effect of combined influence of high temperature, friction and corrosive gas or liquid environment (e.g. gases with increased content of sulfur or liquid metal) parallel structural changes of component usually lead to considerable decrease in resistance to abrasion, corrosion, strength and impact resistance. Simultaneously, substantial increase in the speed of corrosion and oxidation is observed. The product of these processes is e.g. scale, which molecules act as an abrasive material. Specific case of high temperature wear is thermal fatigue, which occurs as a result of periodical heating and cooling of component work surface [3]. It is caused by alternate expansion and shrinkage of surface layer that leads after some period of time to matrix of small cracks, that expand throughout the metal due to exploitation load. This type of wear is typical in case of metallurgical rollers, installation for continuous casting, matrix for hot forging, ingot tongs, clinker grinder elements etc.

Erosion wear is caused by periodical influence of gas or liquid stream poured at high pressure on work surface. The existence of fine molecules of dust and other abrasive in the gas or liquid intensify the erosion process, just as the increase of temperature and corrosive environment. This type of wear is typical for turbine blades, and material which assure best possible resistance to erosion wear are alloys based on cobalt or ceramics [1, 3].

3 Methodology

The comparative study of inserts for high-pressure die-casting moulds after their regeneration was carried out. Two methods were applied for this process: TIG cladding and laser cladding.

The scope of tests included the study of surface microtopography and material structure on metallographic specimens. In particular, the thickness of the overlay layer and the heat affected zone were examined.

The casting mould is used for the needs of the automotive industry. The foundry performs an average of 400 cast per shift, which directly affects the wear process of the mould and its components (inserts). The mould erosion process occurs at the contact boundary between the high pressure of AlSi9Mg aluminum alloy and the mould made of Uddeholm Orvar Supreme—1.2344 (the chemical composition of the mould material is given in Table 1). This material is widely used in die-casting and high-pressure die-casting moulds in the world.

Table 1. The chemical composition of the Uddeholm Orvar Supreme mould material (1.2344).

C/%	Si/%	Mn/%	Cr/%	Mo/%	V/%	Fe/%
0.39	1.0	0.4	5.2	1.4	0.9	bal

For the TIG cladding process, a welding rod made of 45355 W steel with a diameter of 2 mm was used as a filler material. Its chemical composition is presented in Table 2. However, in the laser cladding process, DIM L-13 steel was used (chemical composition is presented in Table 3).

Table 2. Welding rod chemical composition in the TIG cladding process—steel 45355 W.

C/%	Si/%	Mn/%	P/%	S/%	Cr/%	Mo/%	Ni/%	Al/%	Ti/%	Co/%	Fe/%
0.005	0.05	0.02	0.002	0.001	0.06	4.03	18.17	0.136	1.64	12.12	63.5

Table 3. Chemical composition of the welding rod for the laser cladding process—DIM L-13 steel.

C/%	Si/%	Mn/%	Cr/%	Mo/%	Ti/%	Fe/%
0.25	0.5	0.7	5.0	4.0	0.6	Bal

For the TIG cladding process, a Castolin TIG 2202 device was used with the following welding parameters: current intensity $I = 40$A, arc voltage $U = 24$V, tungsten electrode diameter Ø3.2 mm, inert gas shield (argon). The 2 mm diameter, steel wire 45355 W was used as filler material.

On the other hand, for the laser cladding process, a Freden device with a diode laser was used. The stand is shown in Fig. 1. The following parameter values were used in the cladding process: pulse frequency $f = 4.0$ Hz, pulse duration $t = 3.0$ ms, argon protective gas, and gas supply pressure $p_{\mathrm{A}} = 10$ bar. Distance of the welded object from the eyepiece was set as $h_{\mathrm{p}} = 90$ mm.

4 Results and Analysis

The 3D analysis of the surface topography was carried out for the tested samples. TIG cladding method results are shown in the Fig. 2, and comparatively laser cladding results are shown in the Fig. 3. The parameter values of the obtained topography are presented below the graphs, respectively. The tests were carried out in accordance with ISO 25178. The determined parameters were defined as follows: S_a—the arithmetic mean of the

Fig. 1. Laser surfacing station made by Freden company.

surface height, S_q—the mean square of the surface roughness, S_p—the height of the largest peak on the surface, S_z—the maximum surface height, S_v—parameter as the maximum cavity area.

As a result of the observation of the deburring processes, we can conclude that irregularly arranged elements in completely random directions are visible on the surface of the castings' mantle. This can be seen with visual testing (VT), i.e. with the naked eye or with a magnifying glass

where:

- S_a Arithmetic mean deviation of the surface unevenness height from the reference plane.
- S_q Mean square deviation of the surface unevenness height from the reference plane.
- S_p Height of the highest peak on the surface.
- S_z Maximum surface height.
- S_v Maximum cavity area.
- S_t Total height of the surface profile.

There is the significant difference concerning the overly layer size and thickness of heat-affected zone between TIG welding (Fig. 4a) and laser cladding (Fig. 4b) in favor of this second, which is multiple lesser.

The individual surface microtopographies of the samples cladding with the TIG method (Fig. 2) and laser cladding (Fig. 3) differ in the height and density of the peaks and the degree of surface development. The individual values of the parameters for both tested variants differ significantly in favor of the laser-cladding. The relative difference

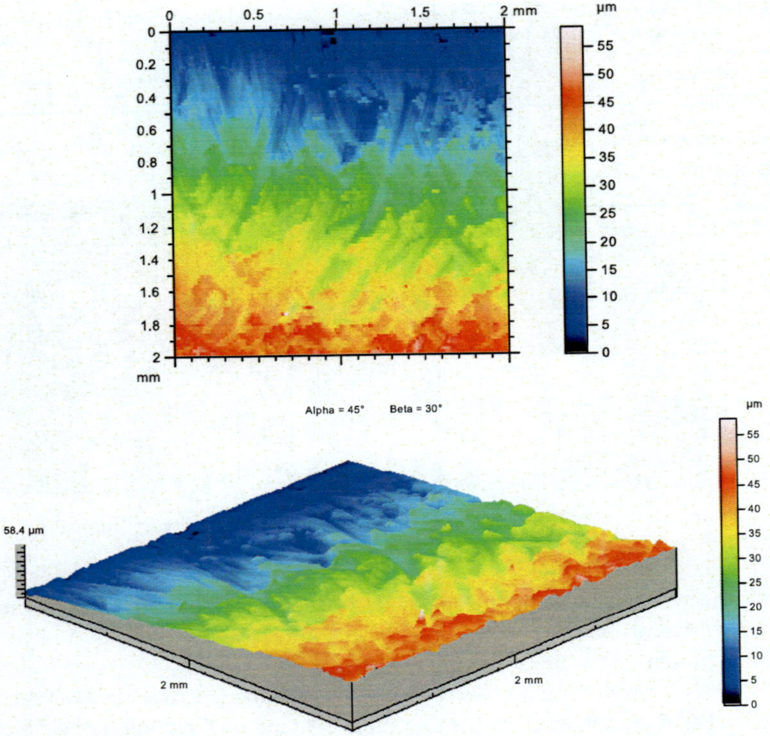

Fig. 2. Surface microtopography for the insert cladding welded with the TIG method. The following parameter values were obtained: $S_a = 10.4$ µm, $S_q = 11.9$ µm, $S_p = 34.3$ µm, $S_v = 24.1$ µm, $S_t = 58.4$ µm, $S_{sk} = 0.01$, $S_{ku} = 1.8$, $S_z = 22.4$ µm.

for the tested parameters was: 45% for S_a (arithmetic mean deviation of the surface unevenness height from the reference plane), 36.8% for S_q (mean square deviation of the surface unevenness height from the reference plane, 56.6% for S_p (height of the highest peak on the surface), 26.6% for S_z (maximum surface height), 15.3% for S_v (maximum cavity area) and 36.4% for S_t (total height of the surface profile). Laser cladding in relation to the TIG process clearly indicates that the more developed surface will be more resistant to the influence of a stream of liquid alloy, which will increase the life of the inserts in permanent casting moulds.

5 Conclusions

The main conclusions from the conducted research are as follows:

- There is a correlation between the roughness of a TIG welding surface and the roughness of a laser cladding.

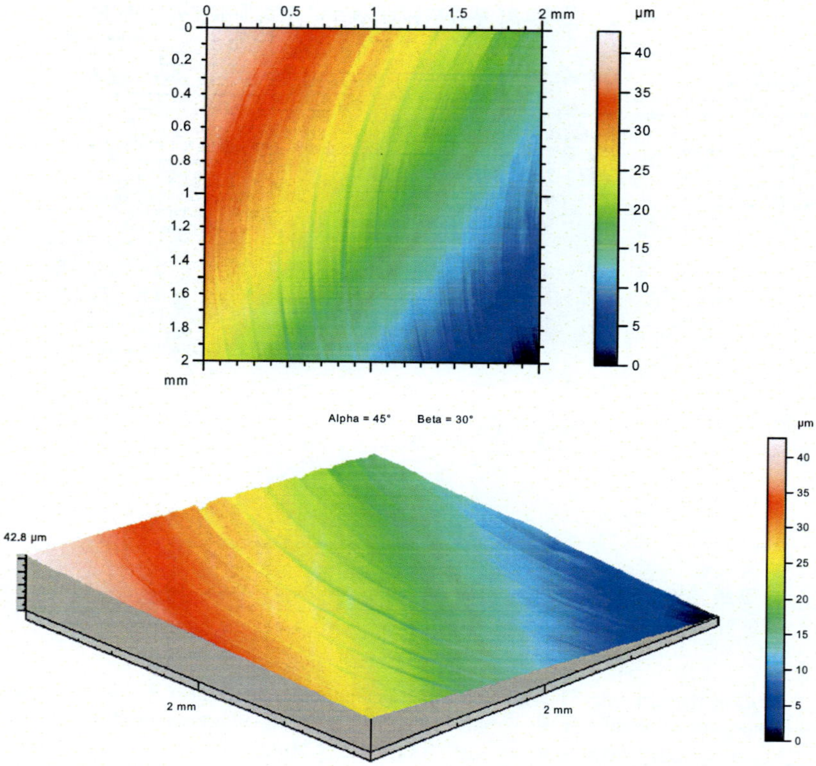

Fig. 3. Surface microtopography for the insert cladding laser method. The following parameter values were obtained. $S_a = 7.17$ µm, $S_q = 8.7$ µm, $S_p = 21.9$ µm, $S_v = 20.9$ µm, $S_t = 42.8$ µm, $S_{sk} = 0.0164$, $S_{ku} = 2.34$, $S_z = 17.7$ µm.

- There is the significant difference concerning the overly layer size and thickness of heat-affected zone in favor of laser cladding.
- Due to the quality of a surface, it is most reasonable to use laser cladding method than TIG welding. The use of this method allowed for obtaining the lowest values of Sa parameters of the sample surface.

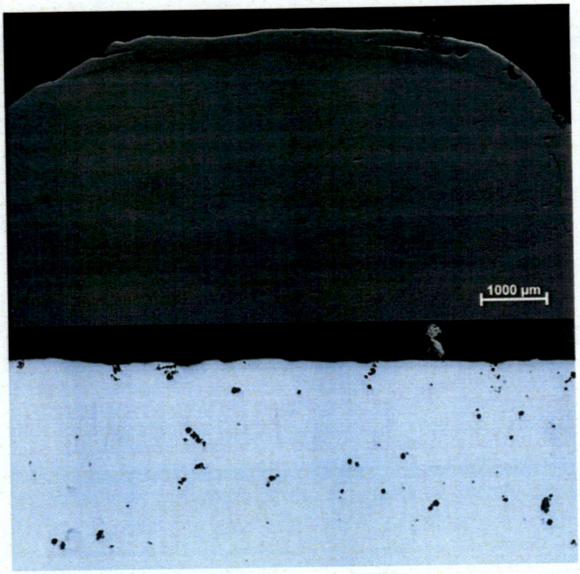

Fig. 4. Microstructure of samples: **a** TIG cladding, **b** laser cladding (common scale for both microstructures).

Acknowledgment. The research was financed under the project Actions 1.2 "Strengthening the innovation potential of Wielkopolska enterprises" of the Wielkopolska Regional Operational Program for 2014–2020 project no. RPWP.01.02.00–30-0080/17: Development of R&D in the GTS enterprise Grzegorz Stawiński as a chance for implementation innovation.

References

1. Mathia, T.G., Pawlus, P., Wieczorowski M.: Recent trends in surface metrology. Wear **271**, 494–508 (2011)
2. Jiang, X.J., Whitehouse, D.V.: Technological shifts in surface metrology. CIRP-Manuf. Technol. **61**/1, 815–836 (2012)
3. Leach, R. (ed.): Characterization of a real surface texture. Springer, Berlin (2013)
4. Whitehouse, D.J.: Function maps and the role of surfaces. Int. J. Mach. Tools Manuf. **41**, 1847–1861 (2001)
5. Novocic, D., Dewes, R.C., Aspinwall, D.K., Voice, W., Bowen, P.: The effect of machined topography and integrity on fatigue life. Int. J. Mach. Tools Manuf. **44**, 125–134 (2004)
6. Stout, K.J., Blunt, L.: Three Dimensional Surface Topography. Penton Press, London (2000)
7. Kubiak, K.J., Liskiewicz, T.W., Mathia, T.G.: Surface morphology in engineering applications: influence of roughness on sliding and wear in dry fretting. Tribol. Int. **44**, 1427–1432 (2011)
8. Sing, R., Melkote, S.N., Hashimoto, F.: Frictional response of precision finished surfaces in pure sliding. Wear **258**, 1500–1509 (2005)
9. Tayebi, N., Polycarpou, A.A.: Modeling the effect of skewness and kurtosis on the static coefficient of rough surfaces. Tribol. Int. **37**, 491–505 (2004)

10. Sedlacek, M., Podgornik, B., Vizitin, J.: Influence of surface preparation on roughness parameters, friction and wear. Wear **266**, 482–487 (2009)
11. Lopatková, M., Bárta, J., Marônek, M., Šugra, F., Kritikos, M., Samardžić, I., Marić, D.: The influence of surface roughness on laser beam welding of aluminium alloys. Technical Gaz. **28**(3), 934–938 (2021). https://doi.org/10.17559/TV-20201102100726
12. Wang, T., Li, Y., Liu, J., Qin, L., Wang, N., Zhang, L., Wang, H., Li, Z.: Milling force and surface topography of Ti-6Al-4V titanium alloy cladded by the laser. Surf. Rev. Lett. **26**(8), 1850185 (1–11) (2018). https://doi.org/10.1142/S0218625X18501858

Estimation of Welding Costs During the Production of the Lower Pedestal Central Ring of the Hydraulic Excavator

Sara Radojicic[1](✉), Pejo Konjatic[1], Marko Katinic[1], Josip Kacmarcik[2], and Andrijana Milinović[1]

[1] Mechanical Engineering Faculty, University of Slavonski Brod, Slavonski Brod, Croatia
sradojicic@unisb.hr

[2] Mechanical Engineering Faculty, University of Zenica, Zenica, Bosnia and Herzegovina

Abstract. The subject of this paper is the calculation of welding costs of the lower pedestal central ring of the hydraulic excavator. Purpose, applications and the main parts and functions of the hydraulic excavator are described in this paper. The effects of major welding methods on the overall standards of time and the main welding costs of the central ring of the lower base of a hydraulic excavator are analyzed and based on the cost calculations of welding, optimal welding procedures are selected for described welds to achieve minimum time needed with the lowest welding cost.

Keywords: Central ring · Hydraulic excavator · Welding · Maintenance

1 Introduction

Only with the use of construction machinery is it possible to carry out large construction projects in a very short time, to achieve the required quality and economic efficiency, and at the same time to humanize human labor. Construction machinery is therefore of great importance to the construction industry. Without them, the construction industry could not survive; they are its backbone. From levers, pulleys, wheels and ropes, construction has evolved to the point where its future lies in electronics, remote control, automata and robots. Therefore, the use of construction machinery requires a thorough knowledge of their capabilities and the areas in which the work is performed, which can be divided into 3 groups: [1].

1. Civil engineering works,
2. Construction works,
3. Other construction works.

Basic operating conditions of any construction machine are working temperature, stress and medium. According to the conditions of exploitation, the parts of the hydraulic excavator are subjected to bending as well as tensile, compressive, impact and dynamic

D. Blažević et al. (Eds.): OTO 2022, LNNS 592, pp. 160–168, 2023.
https://doi.org/10.1007/978-3-031-21429-5_15

stresses. In such conditions, excessive stresses are possible, resulting in unwanted defects. In order to prevent this from happening, it is necessary to carry out visual inspections of possible damages during exploitation and to remove them in a timely manner. In the case of a hydraulic excavator, a hydraulic drive is used. It is a drive based on the transmission of force by liquid, usually mineral oil containing zinc. Contact of the medium with water and impurities in storage should be avoided, and filtering during storage and immediately before use is recommended [2].

2 Hydraulic Excavator

Hydraulic excavators are construction machines used for excavation and demolition work. They have hydraulic drive, which is easier to handle and replaces cable excavators, which have mechanical elements such as winch drive and gearbox. Their range of application is wide, and they are mainly used for excavation. They are also used for demolition work, as even the smallest models are capable of demolishing buildings. Hydraulic excavators are also used to move large amounts of material from one place to another. They are completely dependent on the hydraulic drive [2]. The base consists of a track or wheel that allows the machine to move while supporting the undercarriage. The operator controls the hydraulic excavator with levers to raise and lower the two-piece arm and control the excavator arm. The drive section is located on the undercarriage and can be rotated 360 °. The two-piece arm and the excavator arm are connected to the drive section but are independent during control. Mining and big infrastructure projects are fields where hydraulic excavators are used extensively. These machines generally use tracked walking systems that have high ground adhesion in order to be reliable in the surroundings with harsh ground [3].

2.1 Basic Information About Welded Construction

Each hydraulic excavator consists of parts necessary for its operation. As a representative example, the RKE-2600 excavator model was selected, the main components of which are shown in Fig. 1.

The central ring of the hydraulic excavator was chosen as a representative example of welded construction. For the selected design, an analysis of the main welding costs was performed, and the most favorable welding procedures for each weld were selected. The central ring is the connecting part between the upper and lower parts of the hydraulic excavator and allows the upper part to rotate relative to the lower part of the excavator, which increases the range of motion of the excavator. Since the ring is subjected to extremely dynamic shock and pressure loads, high operational reliability is required, which is achieved through high-quality manufacturing and inspection. Figure 2 shows the central ring of the lower base of the Render RKE Hydraulic Excavator 2600.

2.2 Sequence of Production and Control Activities

The sequence of production and control activities in the manufacture of the product, from the creation of the individual product positions to their assembly into the finished

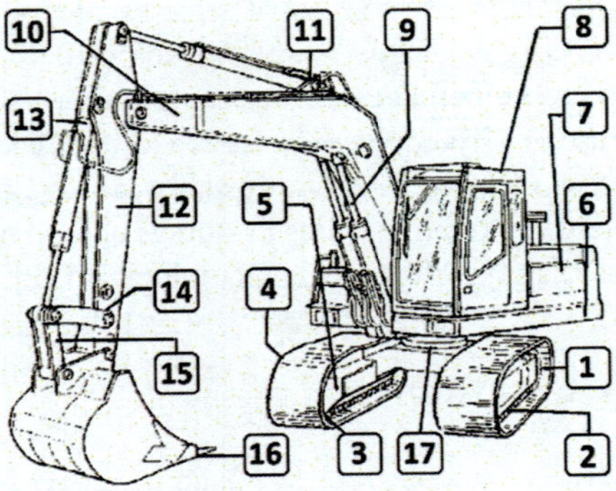

Fig. 1. Main parts of hydraulic excavator RKE-2600 [4] (1—track drive wheel, 2—track guide rollers, 3—track tension wheel, 4—tracks, 5—hydraulic motor, 6—superstructure, 7—covers with sound protection, 8—cabin with control devices, 9—lifting cylinder, 10—excavator boom (boom), 11—cylinder for penetration, 12—grab holder, 13—grab cylinder, 14—grab lever, 15—connecting rod, 16—gripper, 17—main bearing)

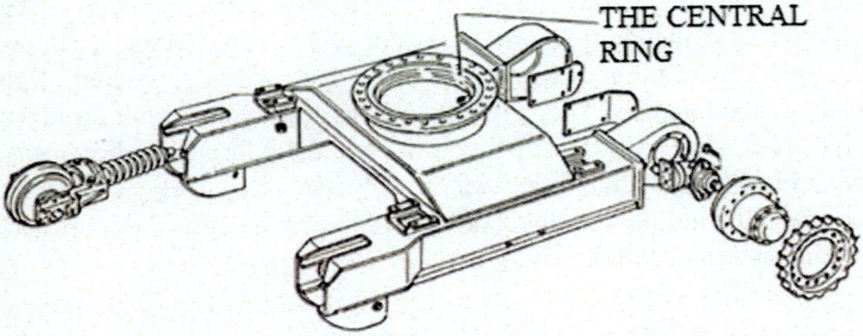

Fig. 2. The central ring of the hydraulic excavator [4]

product, is shown schematically in Fig. 3. The assembly consists of three positions. Each of the positions is made from a plate with certain dimensions (starting material). The starting material is formed into positions by cutting and bending processes, and the positions are assembled into an assembly by welding. Between the individual production steps, a position or an assembly passes through various controls (visual, dimensional, etc.). The control ensures the quality of production and the required reliability of the assembly. Positions of the central ring are shown in Fig. 4.

3 Analysis of the Main Costs of Welding Procedures

Procedure of joining several structural elements, known as welding, is becoming one of the most widely used procedure in the construction, railroad and automotive industries, as well as in shipbuilding and the offshore industry [6]. The definition of the ideal weld is very complex, but it can be said that the joined parts are in continuity. In practice, it is impossible to achieve an ideal weld, but there are several welding procedures that are

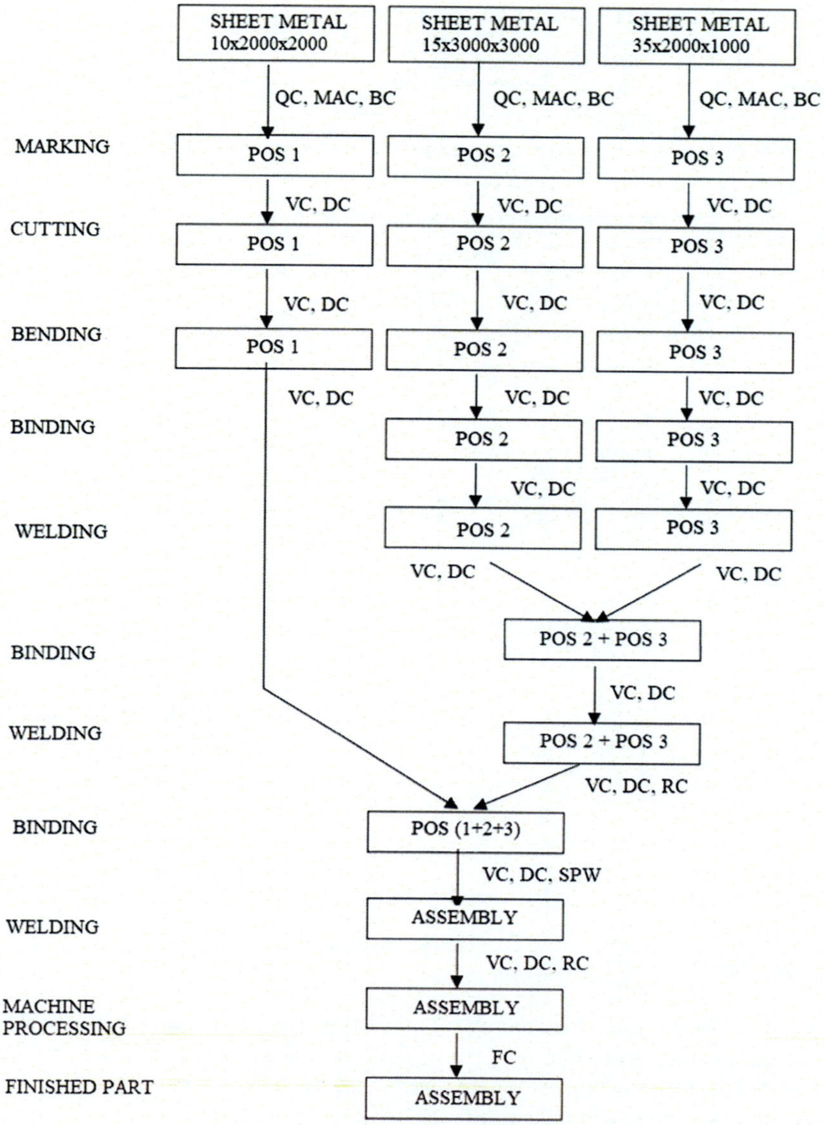

Fig. 3. Sequence of production and control activities [5] (VC—visual control, DC—dimensional control, RC—radiographic control, QC—control quality, MAC—material attestation control, BC—batch control, SPW—stoppage point works controls, FC—final control)

designed to produce acceptable welds. Since there are different types of metals, joints, and applications, the welding engineer must identify the requirements needed for each weld and select the appropriate welding procedure [7].

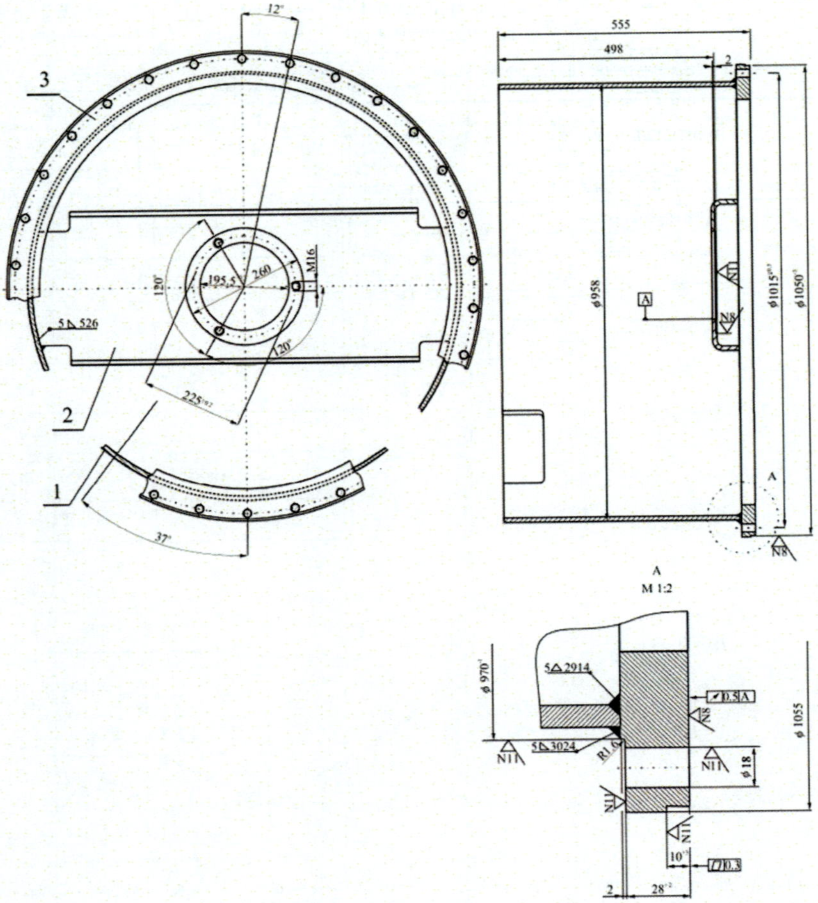

Fig. 4. Positions of the assembly

The purpose of technological analysis and cost optimization is to achieve maximum profit. How to ensure maximum profit, it is necessary to choose the optimal solution for the welding process, optimal groove shape and appropriate groove tolerances. Profit can be expressed as the difference between the price that the product can achieve in the market and procurement and manufacturing costs. As the price of the welded product is defined by the market, it is desirable to reduce production costs to the smallest possible measure to maximize profit. The correct choice of welding method and of other factors that affect the cost of the welded product can be achieved the smallest costs, and thus the highest profit. The control and measurement of all variables of the process are the most important steps in the optimization of all welding parameters involved [8]. The most common electric arc welding procedures are:

MMA Welding—Manual Metal Arc Welding,

MIG/MAG Welding—Metal Inert/Active Gas Welding,

SAW Welding—Submerged Arc Welding,

TIG Welding—Tungsten Inert Gas Welding.

MAG, MIG and TIG welding belong to semi-automated, while SAW welding belongs fully automated welding procedures. Each of these procedures is performed differently and their main welding costs are also different. Increased attention must be paid to the strength calculation, the composition of the material used and welding to avoid any failures [9]. It is very important to accurately determine the residual stresses, since their occurrence is one of the consequences of the process of welding. Nowadays it is indispensable to perform numerical simulations of the process of welding with FEM [10–12]. In the following chapters, the main costs of individual welding procedures are briefly described.

3.1 Manual Metal Arc Welding (MMA Welding)

The main costs of MMA welding are:

1. Electrode cost,
2. Power cost,
3. Personal income cost (workers' wages),
4. Costs of power source (machine) for welding.

The cost can be expressed in HRK/kg deposit or HRK/m weld length. It is argued that it is more appropriate to express the cost in HRK/kg deposit. This is because a different amount of deposit can be used for 1 m of weld, and the cost is proportional to the mass of the deposit.

Electrode cost can be estimated from the following expression: [13]

$$T_{\text{electrode}} = C^{1}_{\text{electrode}} \times k'_{t} \tag{1}$$

where $C^{1}_{electrode}$ is the unit price of electrodes and k'_t is the electrode melting coefficient, which expresses the amount of electrode mass together with coating required to deposit 1 kg of deposit. This coefficient depends on the thickness of the electrode coating, the amount of iron powder added to the coating to increase its effectiveness and the amount of waste ("stick") from the electrode that the welder leaves unused. This waste should be as small as possible and can be 30–50 mm if care is taken. A larger waste due to carelessness causes higher costs.

The expression for power cost is equal for every welding process: [13]

$$T_{\text{electricity}} = \left[\frac{U \times I}{1000 \times \eta_{\text{s}}} \times \varepsilon + N_0(1 - \varepsilon)\right] \times \frac{1}{k_{\text{t}} \times \varepsilon} \times C^{1}_{\text{electricity}} \tag{2}$$

where U is the arc voltage, I is the welding current, η_s is the degree of useful operation of the machine and ε is intermittency, the time of switching on the electric arc. N_0 is the power consumed by the machine at idle when the arc is not burning. Power is then used for the working fan, friction, magnetic field dissipation, and heating of the conductors in the machine. Also k_t is the electrode melting coefficient and $C^1_{\text{electricity}}$ is the unit price for current.

Personal income cost is also equal for every welding process and it can be calculated from the following expression: [13]

$$T_{\text{PIE}} = \frac{PIE}{k_t \times \varepsilon} \tag{3}$$

where PIE is the gross amount of PI (personal income) that results when the net PI is added to the compulsory contributions to the social community (pension and health insurance and other funds).

The last main cost is the cost of the welding machine for which the following expression applies: [13]

$$T_{\text{machine}} = \frac{C_N}{\text{number of working hours per year}} \times \frac{1}{k_t \times \varepsilon} \tag{4}$$

where C_N is the purchase price of the machine. This includes the annual depreciation rate, the annual amount for maintenance, and the interest or tax on the operating fund. The number of working hours per year of the machine in one or more shifts depends on the actual use of the machine. The total main cost is equal to the sum of all individual welding costs.

3.2 Metal Inert/Active Gas Welding (MIG/MAG Welding)

The approach to calculate the main cost of MAG/MIG welding is similar to the approach to calculate the main cost of MMA welding. The difference with the main cost of MMA welding is that the cost of the wire used instead of the electrode is added to the cost of the welding gas. The parameters of MAG /MIG welding also differ significantly from the parameters of the MMA process. Gas cost can be calculated from the following equation: [13]

$$T_{\text{CO2}} = C_N \times K_{\text{CO2}} \tag{5}$$

where T_{CO2} is CO_2 gas cost and K_{CO2} is CO_2 consumption coefficient.

3.3 Submerged Arc Welding (SAW)

The calculation of the main cost of SAW welding is similar to the calculation of the same with the MMA and MAG/MIG process. The difference with the MAG/MIG process is that powder is used instead of a shielding gas, while the SAW process uses wire instead of the electrode used in the MMA process. Approximately 1,2 kg of powder is required for 1 kg of deposition. The exact expression for the cost of powder is following: [13]

$$T_p = 1,2 \times C_p \tag{6}$$

where T_p is powder costs and C_p is powder price.

3.4 Tungsten Inert Gas Welding (TIG Welding)

The calculation of the main costs of TIG welding is analogous to the calculation of the main costs of previous welding processes.

4 Welding Costs of the Selected Representative

In the production of any product, the goal is to obtain the highest possible profit. It will depend primarily on the choice of the welding process, the choice of the groove and the appropriate tolerances of the groove. It can be expressed as the difference between the price of the product on the market and the total cost. Since the price of the product is determined by the market, the cost of production must be reduced in order to maximize the profit. When welding the central ring of the lower part of the base of the hydraulic excavator, due to the thickness, shape of the groove and inaccessibility, not all welds can be welded using the same procedures. The most important thing is to choose a technological welding process so that the cost is minimal. Therefore, the welding of weld 1 connecting Position 1 is performed using the SAW process, and the other three welds are welded using the MAG/MIG process. The total main costs of each weld and the whole series are shown in Fig. 5.

WELDING PROCEDURE	WELD 1		WELD 2		WELD 3		WELD 4	
	Total main costs T_z HRK	Total main costs for the entire series T_{ser} HRK	Total main costs T_z HRK	Total main costs for the entire series T_{ser} HRK	Total main costs T_z HRK	Total main costs for the entire series T_{ser} HRK	Total main costs T_z HRK	Total main costs for the entire series T_{ser} HRK
REL	27,71	1385,5	6,80	340	20,39	1020	50,81	2541
MAG/MIG	12,49	614,5	3,06	151	9,19	452,5	22,90	1127
SAW	8,09	404,5	-	-	-	-	-	-

Fig. 5. Total main costs for individual welds and the entire series

By choosing a technological welding process for each weld, the lowest welding cost for the entire assembly is achieved. Based on the values already calculated, the main welding cost for the entire series of 50 pieces is T_{ser} = HRK 2134.7. If all welds were welded using the MMA process, the cost for the series would be $T_{ser} = 5286.5$ HRK, while welding all welds using the MAG/MIG process would reduce the cost for the series to $T_{ser} = 2345$ HRK. MAG/MIG welding results in a lower amount of main costs compared to REL welding, but still a higher amount than the already chosen technological solution (a combination of SAW and MAG/MIG welding).

5 Conclusions

This paper describes importance and use of construction machinery including hydraulic excavators which are commonly used as construction machinery to carry out larger construction projects to reduce construction time.

As a representative example for analysis of welding cost in construction machinery production, a central ring of one hydraulic excavator was chosen. In order to select optimal welding procedure regarding production cost, a cost of MMA Welding, MIG/MAG Welding, SAW Welding and TIG Welding are calculated and analyzed for four welds of the central ring of excavator. Analysis was performed for one central ring as well as for production series of 50 pieces.

Analyzing the main costs of welding the central ring of the hydraulic excavator, it was found that the combination of the SAW and MAG/MIG processes is a cost effective solution in this case. With the proposed welding methods, a full penetration of the weld root also have to be achieved in order to avoid the possibility of crack formations and weld failure.

References

1. Bucksch, H.: Dictionary of Civil Engineering and Construction Machinery and Equipment. Bauverl (1971)
2. Sinclair B.: Hydraulic Excavators. Japonica Press (2006)
3. Dong, Z., Quan, L., Yang, J.: Tracked walking mechanism for large hydraulic excavators. Autom. Constr. **96**, 88–102 (2018)
4. Đuro Đaković Specijalna Vozila: Instructions for Handling and Maintenance of RKE 2600 Excavator. Slavonski Brod (2003)
5. Radojicic, S.: Production of the central ring of the lower base of the hydraulic excavator. In: Mechanical Engineering Faculty in Slavonski Brod, Bachelor's Degree (in Croatian) (2016)
6. Maksimovic, M., Maksimovic, K., Stamenkovic, D., Vasovic Maksimovic, I.: Initial fatigue life estimation of welded structural components. Tehnički Vjesn **28**(4), 1099–1104 (2021). https://doi.org/10.17559/TV-20200414015501
7. Houldcroft, P.T.: Welding Process Technology. Cambridge University Press, Cambridge (1977)
8. Edwin, R.D.J., Jenkins, H.D.S.: A review on optimization of welding process. Procedia Eng. **38**, 544–554 (2012)
9. Mohyla, P., Havelka, L. i Kouřil, K.: Mechanical properties testing of P92 welded joints prepared by manual metal arc welding. Tehnički Vjesn **25**(1), 60–63 (2018)
10. Deng, D., Murakawa, H.: Numerical simulation of temperature field and residual stresses in multi-pass welds in stainless steel pipe and comparison with experimental measurements. Computat. Mater. Sci. **37**, 269–277 (2006). https://doi.org/10.1016/j.commatsci.2005.07.007
11. Stamenkovic, D., Vasovic, I.: Finite element analysis of residual stress in Butt Welding two similar plates. Sci. Technical Rev. **59**(1), 57–60 (2009)
12. Bonifaz, E.A.: Finite element analysis of heat flow in single-pass arc welds. Weld. Residual Suppl. 121–125 (2000)
13. Samardžić I., Klarić, Š.: Analysis of welded structures, (in Croatian), Digital Textbook

Automatic Control of Recirculation System for Respiratory Control

Miloš Jovanović[1], Djordje Dihovicni[2(✉)], and Muharrem Hilmi Aksoy[3]

[1] Independent Designer and Contractor of Technical Projects, Belgrade, Serbia
[2] Department of Computer-Mechanical Engineering, Academy of Technical Applied Studies Belgrade, Belgrade, Serbia
dihovicni@visokatehnicka.edu.rs
[3] Department of Mechanical Engineering, Konya Technical University, Konya, Turkey

Abstract. The issue of providing fresh air and technologies for its maintenance within physiologically acceptable references represents an active field of innovative engineering in the field of medicine, industry, army, space program, sports and lately and everyday life of people in cities. The problem-solving approach is generally divided into the construction of personal devices, for living in environments with reduced oxygen concentration and/or increased concentration of harmful elements that would lead to disruption of vital functions and system solutions for air recovery and ventilation in closed indoor units. Although 400 years have passed since the first technical concept, this area is still a challenge to optimize systems and devices and enable people to realize their activities in all potentially and real risk areas, in terms of maintaining respiratory function and metabolism without negative effects and increased efforts as close as possible to the stay in the natural environment for which man is prepared with his biological apparel.

Keywords: Industrial plants · Automation · Recirculation system

1 Introduction

The paper deals with the problems of indoor air treatment without the possibility of establishing ventilation with fresh air, and the design and management of this very responsible technical system. The very problem of respiratory treatment is an important part of many industrial and medical processes [1], but it is also becoming a growing problem of general pollution and the need to perform them in adequate and controlled conditions of daily activities [2]. The problem is caused by air contamination or a disorder air composition and inadequate oxygen percentage.

It is realistic to expect such a situation in industry, especially mining environment or in zones of military action, or in an area affected by natural disasters [3], that have caused the environment to be unfavorable in terms of maintenance breathing functions, and etc. [4, 5]. Also, conditions in which precise maintenance is required, the composition and purity of the respiratory medium may be necessary in high-techs stems and laboratories, the so-called clean rooms, medical systems and in any possible situation

D. Blažević et al. (Eds.): OTO 2022, LNNS 592, pp. 169–175, 2023.
https://doi.org/10.1007/978-3-031-21429-5_16

when it is necessary to perform treatment breathing medium, without the possibility of establishing ventilation/recuperation using external "natural" air mass [6–8]. Each of the possible scenarios requires installation of a system that would regulate the respiratory medium. It is done through work engineering of a mobile autonomous container-type system, with the aim establishing full automation of work [9], improving air treatment and by optimizing parameters in real time. The purpose fullness of the project is reflected involving the problem of treating air that is in a closed loop inside premises [10]. Regulating and synchronizing the work of all elements is done by continuous monitoring and treatment of the respiratory medium [11], to ensure physiological acceptable air and direct optimization of the entire autonomy of the system [12]. In that manner it provides greater autonomy of living within an isolated space, accommodation sensitive equipment and/or materials and complete separation from impacts of external factors.

Combining theoretical and practical knowledge in the field of automation, mechanical constructions, thermotechnical and ventilation systems, with application standards and regulations is systematically improved the management of this responsible technical system.

2 Realization of a Physical Model

For the realization, it is necessary to look at the existing solutions in the context of getting to know each other with already proven practice. The existing solutions that were selected served as reference points during the examination of the functionality of the obtained results and perceiving the success of the project. In accordance with the goal of this project [13], all calculations and simulations developed in adequate software making it significant, saves the time needed to obtain a finally acceptable solution [14–16]. Forming physical model, based on the results obtained in a further iteration, with possibly the necessary recalculations and simulations of the system operation can be reached realization and selection of an acceptable solution.

As part of the project, a physical model is implemented. Based on the obtained budget values selected techno-economically and most suitable elements of the system are chosen, in order to form a realistic technical system. It is realized as well as in the budget part, with paying attention to the concrete work of the system without aesthetic ones details that is negligible for this phase of the project. The assumption is that it will be possible to use ready-made ones elements that are on the market, to which adjustments will eventually be made obtained by values which are strived for in the ideal performance. The purpose of forming the physical model is a check of functionality as well as compliance with realistic possibilities for further serial production of the modular system.

Basically, this technical system is a respiratory machine which supports the breathing process and performs the treatment of the respiratory medium, in order to maintain it parameters that affect vital functions. In practice, often deal with various gas mixtures, which are used according to the purpose and conditions in which the activities will take place. Our setting is already indicated in the introduction, in which it is stated that the system is designed by an isolated spatial unit (which is not structurally the subject of this work) which envisages the stay of human staff at a certain time period in which the technical system inside the room is constantly monitored and performs all necessary actions in order to maintain the conditions in the reference values most suitable for living from the aspect of primarily health, but also comfort to reduce the stress caused by staying in complete isolation. When the comfort is mentioned, a focus is on the biological characteristics of the environment in context, above all, breathing and respiratory support as well as overall ambient conditions that directly affect the physiological processes that the body performs to maintain its integrity.

By breathing, a person ventilates the air when he brings into body new ones amounts of oxygen. At the same time, breathing removes the products from the body carbon dioxide, formed as a product of metabolic work. These two most important functions which are realized in each respiratory cycle are the basis for further developing the structure and determining the critical points of work of the designed technical system. The work of the technical, mechanical apparatus is in direct interference with the biological apparatus of man, hence we speak of very responsible engineering, and a whole process can be viewed as the mutual work of two subsystems that interact, communicate through the exchange of a common gas medium. The very fact is that the space, in which the person resides, is isolated from the outside ambience, and fresh air mass is a technical problem that will be successful solved by applying the known technological processes of respiratory medium treatment, through removing unwanted carbon dioxide and enriching the air as needed the amount of oxygen. In this way, the air is maintained in its biological borders and achieves the full effect of the system. This system of work implies that the gaseous medium remains in a continuous stream, without mixing with new amounts of air from the natural environment, which is a closed loop in which everything is done by necessary activities called recirculation.

However, in order to achieve all the necessary characteristics, air needs to be treated and other conditions in which the process takes place so that the end result is possible at all that is, in order to fully achieve positive results. On the structure air and ambient conditions that are directly related to its quality, factors other than composition also affect. These are air temperature, humidity and speed movements, and gas flow through the system and inside the room.

Chemically, air is a mixture of persistent gases, Nitrogen (N_2), Oxygen (O_2), Argon (A_r) and Carbon Dioxide (CO_2) which together make up 99.99% air composition. The composition of the air includes moisture that has a variable value of which is influenced by many factors.

The percentage of moisture is in all technological processes as well as in the treatment of living space must be taken into account, because, it leads to very large change in system state.

Table 1 provides an overview of the composition of pure dry matter air:

Basically, this technical system is a respiratory machine that supports the breathing process and at the same time performs the treatment of the respiratory medium, in order to maintain the parameters that affect vital functions.

Display of recirculation within an isolated spatial unit is shown in Fig. 1.

The complete process takes place in a closed container-type facility. The choice is derived from the usable characteristics that this type of object has in mind possibilities

Table 1. Overview of the composition of pure dry matter air.

Name of the element	Chemical label/formula	Mass share in %	Volume share in %
Oxygen	O_2	23.01	20.93
Nitrogen	N_2	75.31	78.10
Argon	Ar	1.286	0.9325
Carbon dioxide	CO_2	0.04	0.03
Hydrogen	H_2	0.001	0.01
Neon	Ne	0.0012	0.0018
Helium	He	0.00007	0.0005
Krypton	Kr	0.0003	0.0001
Xenon	Xe	0.00004	0.000009

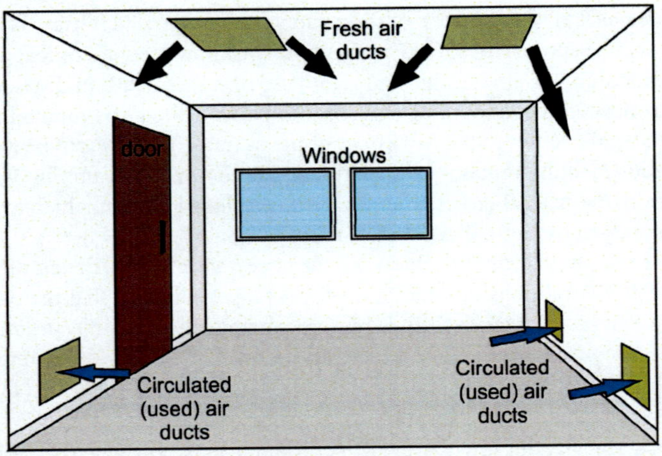

Fig. 1. Display of recirculation within an isolated spatial unit.

of relatively easy transport and relocation as needed. Also, another one advantage over a fixed solid object is a system of modularity that container facility allows.

By connecting multiple container objects it is possible to form a complex that meets the needs of the field in the given conditions. Change scheduling, capacity and equipment, make this system a realistic choice for many activities in which the isolated space could be integrated.

In Fig. 2, it is presented the selected model Office container model type-20. This is the standard type most commonly used. Its dimensions match standard transport conditions without the need to engage oversized transportation and additional relocation costs. A manipulation of an object in space can also be done very easily, by standard working machinery such as forklift or crane arm on the truck.

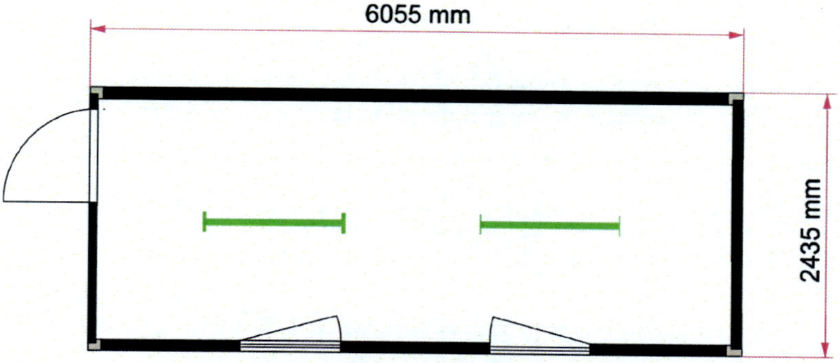

Fig. 2. The model office container model type-20.

To maintain the respiratory medium, in this case air, it is necessary remove emitted carbon dioxide from its composition. Any system that works according to the principle of recirculation has a special part that serves this purpose. Generally it comes down to a canister with an absorbent that reacts with CO_2 and binds it to itself, missing other components such as O_2 and N_2. Canister presents impermeable space through which gas passes in one direction. So the intake air will be under by the influence of the fan operation, to be directed from the suction pipe through the one-way ones valves into the space with the absorbent. A chemical reaction takes place in the canister itself.

Schematic representation of absorbent canisters and reactions with CO_2 is presented (Fig. 3).

3 Description of Mechanical and Biological Air Treatment Systems

Process management and process automation are practically the essence of this project with the described air treatment, without which it would not make sense at all any management. So, the recirculation type air treatment itself is the basis for establishing automation and control processes. Harmonization of all elements is essential for achieving optimal values at all times active phases, i.e. during the stay of the staff in isolation. Continuous work and troubleshooting is a technical requirement that the control system must meet the highest quality. The choice of management method was one of the important phases during realization of this project.

System architecture is the simplest be represented as a network of the following elements:

- assist elements (buttons, thermostats, timers flow velocity, motion detectors) to collect the given information and forward them to the communication line;
- system devices required for system operation (system power supply, line connection devices, etc.).

System design involves optimizing and aligning more parameters in order for the process to take place in the most ideal mode of operation. The structural block diagram

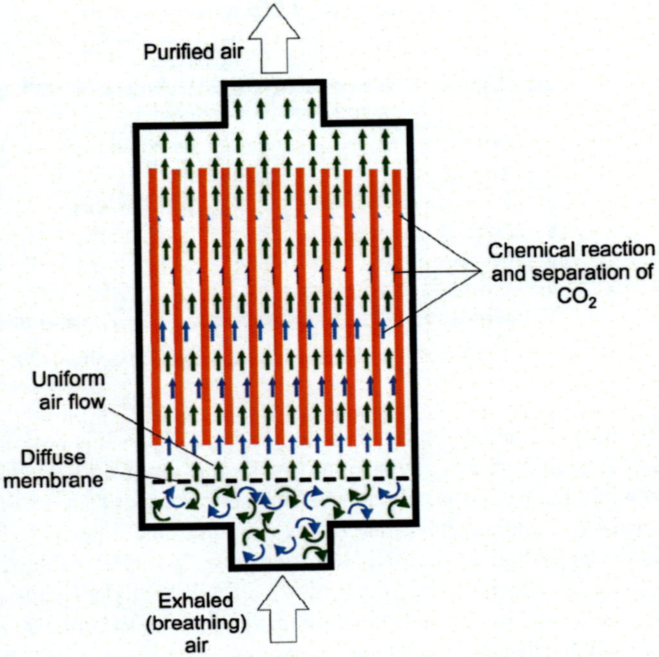

Fig. 3. Schematic representation of absorbent canisters and reactions with CO_2.

allows monitoring the process through subsystems, which can be further defined and analyzed through the introduction of portable function, in order to define functional units that can be later processed by simulating the process and determining in the physical model through the subsystem of the total system functions.

The control module provides information on the gas status in the pressure gauge tank (oxygen cylinder) and based on set reference values reports on the available quantity in the function of maintaining the autonomy of the system. On the alarm indication the required bottle replacement action can be taken in time prolonging the autonomy of being in isolation.

Based on certain functions, we can describe the system through equations and over adequate program to perform the necessary simulations and checks of the system before formation of the physical system. Also, by defining a subsystem, it is now also defined the system as a whole and a realistic scheme of the control unit can be shown.

4 Conclusion

The aim of this paper is the formation of an automatically regulated closed-recirculation system in an isolated one spatial unit, with the task of maintaining the functional parameters of the respiratory medium during the active phase of use.

Based on the input parameters, which concern the projected values for the predicted ones, the maximum number of persons in the spatial module, the required volume exchange of breathing media in a unit of time and initially adopted technical solutions is accessed by constructing a default system and performing initial calculations. Based on the obtained results by comparing with the desired characteristics it is performed further system optimization and structuring.

Successful application of such a system would be found in the industrial environment, medical systems, military-technical systems, but also private housing structures that are designed to stay in limited conditions during the period of threat.

References

1. Zhaoxu, Y., Shugang, L., Hongbin, D.: Razumikhin-Nussbaum-lemma-based adaptive neural control for uncertain stochastic pure-feedback nonlinear systems with time-varying delays. Int. J. Robust Nonlinear Control **23**(11), 1214–1239 (2013)
2. Dihovicni, D., Asonja, A., Radivojevic, N., Cvijanovic, D., Skrbic, S.: Stability issues and program support for time delay systems in state over finite time interval. Phys. A Stat. Mech. Appl. **1**(538), 122815 (2020)
3. Shiquan, Z., Aragona, P., Shining, C.: Advances in Engineering Science and Equipment Engineering II, ICEESE 2016, Conference 2016, vol. 1, pp. 207–211. London (2011)
4. Dihovicni, D.: Decision making and fundamental matrix approach in process safety. Int. J. Comput. Intell. Syst. **6**(4), 658–668 (2013)
5. Bolton, W.: Mechatronics, Electronic Control Systems in Mechanical and Electrical Engineering, 3th edn. Pearson Education Limited, England (1999)
6. Berger, H.: Automating with STEP 7 in LAD and FBD. Publicis MCD Verlag, Erlangen and Munich, SIEMENS (2003)
7. Alciatore, D.G., Histand, M.B.: Introduction to Mechatronics and Measurement Systems, 3rd edn. McGraw Hill, International edition (2007)
8. Blagojević, N., Dihovični, D., Krunić, V.: Analysis of solar plant for the production of hot consumption water for the hospital. Adv. Eng. Lett. **1**(1), 23–27 (2022)
9. Xiuzhi, L., Yaping, D.: Fault tolerant control for fixed set-point control nonlinear networked control systems. Tech. Gaz. **23**(1), 35–41 (2016)
10. Kovačević, Z., Vukašinović, V., Gordić, D.: Piecewise techno-economic analysis of led com bulbs for street lighting. Appl. Eng. Lett. **7**(2), 74–80 (2022)
11. Lima, R., Barros, D.: Identification of time delay systems: a state space realization approach. IFAC Pap. Online **48**(8), 254–259 (2015)
12. Dihovicni, D., Nedic, N.: Simulation, animation and program support for a high performance pneumatic force actuator system. Math. Comput. Model. **48**, 761–768 (2008)
13. Argyll, J., Hill, L.: Concerning the amount of nitrogen gas in the tissues and removal by breathing almost pure oxygen **6I2**(2), 309–321
14. Sriboon, W., Pansila, P., Khunthong, S., Sukprasong, S.: Energy audit and evaluation of specific energy consumption of lighting systems in a new building: a Kasetsart university case study. Appl. Eng. Lett. **6**(4), 157–165 (2020)
15. Yong, L., Yanping, Y., Chaofeng, L., Xu, H., Xiaosong, Z.: Human responses to high air temperature, relative humidity and carbon dioxide concentration in underground refuge chamber. Build. Environ. **131**, 53–62 (2018)
16. Mann, B., Patel, B.: Stability of delay equations written as state space models. J. Vib. Control **16**(8), 1067–1085 (2010)

Draught Beer—Maintaining the Quality Through Hygienic Measures

Kristina Habschied(✉), Snježana Keleković, Ana Domaćinović, Tanja Mađarević Pavetić, and Krešimir Mastanjević

Faculty of Food Technology Osijek, University of J.J. Strossmayer, F. Kuhača 18, 31 000 Osijek, Croatia
khabschi@ptfos.hr

Abstract. Draught beer has always been popular among consumers. The introduction of cooled dispensers made tap beer more desirable. However, since beer is a nutritious beverage and fairly unstable at room temperatures, many undesirable changes may occur. Undesirable changes regard the microbiological stability, colloidal stability, sensory changes, etc. which result in spoiled beer or spoiled aroma. The key is to apply certain hygienic measures and to respect the legislation related to beer quality. To achieve the wanted quality control it is important to apply and provide proper pre-dispensing measures (appropriate temperature for keg storage) and sanitation during dispensing (clean valves, clean pipes, etc.). Aerobic and anaerobic bacteria can cause different changes during storage especially if the beer is not served on a daily basis. Turbidity, color, foam quality, sensory deterioration can occur if proper sanitation measures are not undertaken. The input of efforts and resources involved in draught beer quality control should be continuous.

Keywords: Draught beer · Taps · Sanitation · Dispensers · Quality

1 Introduction

1.1 Draught Beer in Dispensing Systems

Even though beer is considered to be a safe beverage in terms of spoilage microorganisms, it is also a very nutritious and prone to spoilage if not being subjected to proper hygienic measures during and after production, especially draught beer.

The production of beer is a closed system, with small chance of contamination during production. Generally, even though beer contains certain antimicrobial compounds such as hop compounds, ethanol, CO_2 and SO_2, it is a suitable beverage for microbial contamination due to its nutritious nature. Low oxygen levels and the low pH can hinder the microbial proliferation as well. Additional operations such as filtration, low temperature storage, pasteurization prior to packaging in sealed cans, kegs or bottles also interfere the microbiological contamination [1]. Tunnel pasteurization applied after the filling is also an important factor in reducing the microbial contamination.

D. Blažević et al. (Eds.): OTO 2022, LNNS 592, pp. 176–185, 2023.
https://doi.org/10.1007/978-3-031-21429-5_17

However, draught beer and related equipment is another story and can easily be contaminated with different microorganisms, some of which are considered pathogens. Continuous inflow of oxygen, higher temperatures in the dispensing system, non-proper dispensing (touching the fluid in the glass with nozzle, touching the glass with nozzle, etc.) increase the probability of microbial contamination [1, 2].

Hygienic and sanitation measures ensure quality of beer by eliminating and minimizing the risk of contamination. Optimization of cleanings procedure is immense in order to improve hygienic conditions of draught beer dispensers but to reduce the cost of cleaning as well.

2 Draught Beer Contaminants

Quality maintenance must be designated in every bar that serves draught beer and they have to determined by the country's legislation.

Beer contains ethanol and CO_2, two antimicrobial compounds. Nonetheless, nutritive compounds such as proteins, carbohydrates, and numerous other organic compounds contribute to microbial growth. Thus, many microorganisms proliferate in draught beer equipment, colonizing the pipes and fixtures. Filtered and pasteurized beer is less likely to contain yeast and bacteria, but craft beer, commonly not subjected to filtration and pasteurization, contains certain number of viable yeast cells [3]. Together with minerals from beer they deposit in lines and fixtures. Biofilm is in close contact with the "new" beer coming from the keg through the pipe lines and significantly deteriorates the quality of draught beer [3].

2.1 Common Contaminants

Hop compounds, alcohol, CO_2 and SO_2, shortage of oxygen and low pH contribute to beer being somewhat resistant to microbial contamination. Additionally, the applied unit operations (filtration, storage at low temperatures and pasteurization) effectively decrease the probability of contamination.

According to several authors [4–7], microbes in brewing can be brought down to several species many of which can cause spoilage and quality deterioration, but they are not designated as pathogens [7, 8]. According to Back [6] microbial cultures associated with brewing can be subdues to several categories described in the following sections.

Absolute beer spoilage organisms

Mostly resulting in off-flavors and turbidity and tolerating the selective environment in beer, absolute spoilage microorganisms proliferate with short adaptation time. Here are some examples of absolute spoilage microorganisms.

Lactobacillus genera:

Lactobacillus brevis,
L. lindneri,
L. brevisimilis,
L. frigidus,

L. coryniformis,
L. casei,

Pediococcus genera:

Pediococcus damnosus,
Pectinatus cerevisiiphilus,
P. frisingensis,

Megasphaera cerevisiae, Selenomonas lacticifex and *Saccharomyces cerevisiae* (ex. *Diastaticus*) belong to this category as well [6, 9].

Authors [10, 11] determined a previously unknown beer-spoilage *Lactobacillus* sp. Strains. The research conducted by [11] described a Gram-negative, non-motile, strictly anaerobic bacterium with weak beer-spoilage ability. This bacterium showed significant difference than any previously known anaerobic beer-spoilage bacteria involving *Pectinatus* spp., while [12] reported *M. cerevisiae.*

Yeast (inoculant) contaminants are considered to be [13, 14]:

Selenomonas lacticifex,
Zymophilus raffinosivorans.
Zymophilus paucivorans.

The severity of contamination can vary depending on microorganism. On that note *Lactobacillus* spp. Causing beer spoilage often belong to obligate heterofermentative species as:

L. brevis,
L. lindneri.

A strain isolated by [4, 10, 14, 15] can cause strong beer spoilage as well. However, some facultative heterofermentive *Lactobacillus* strains cause weak beer spoilage, according to [6, 10, 11, 16].

Potential beer spoilage organisms

They are not indigenous to beer. However, beers displaying increased pH, lower hop compounds concentration, lower fermentation rate and subsequently lower alcohol content or increased O_2 content present a suitable medium for proliferation of potential beer spoilers. Following microbes adapt to grow in beer [6, 14]:

Lactobacillus plantarum,
Lactococcus lactis,
Lactococcus raffinolactis,
Leuconostoc mesenteroides,
Micrococcus kristinae,
Pediococcus inopinatus,

Zymomonas mobilis,
Zymomonas raffinosivorans.

Indirect beer spoilage organisms

These kind of organisms die out in finished beer. However, they are active and proliferate at different stages of the beer production process. They may cause off-flavors in finished beer. Commonly they are found to contaminate the pitching yeast. According to [6], this category involves different species of enterobacteria and some, mostly wild and aerobic, yeasts. Reportedly *Obesumbacterium proteus* and *Rahnella aquatilis* are recognized as important enterobacterial deterioration microorganisms violating beer production process [17].

Clostridium spp. Originating wort production or brewery adjuncts [18, 19] are considered to be indirect beer spoilage microorganisms.

Wild yeasts mostly cause super-attenuation while beer acquires off-flavors involving esters, fusel alcohols, diacetyl, phenolic compounds and H_2S.

Lactobacillus and *Pediococcus* contamination results in lactic and acetic acids, diacetyl and acetoin off-flavors.

Acetobacter and *Gluconobacter* cause acetic acid off-flavor.

Enterobacteria can slow down or completely halt the fermentation and results in formation of apparent total n-nitroso compounds. This leads to increase levels of dimethyl sulphide, acetaldehyde, fusel alcohols (n-propanol, iso-butanol, iso-pentanol, iso-amylalcohol), vicinal diketones, acetic acid and phenolic compounds.

Zymomonas pollution leads to increased concentrations of H_2S and acetaldehyde.

Off-flavors related with *Pectinatus* spp. Result in H_2S, methyl mercaptane, propionic, acetic, lactic and succinic acids, acetoin in final product.

Contamination with *Megasphaera* commonly leads to H_2S, butyric, valeric, caproic and acetic acids and acetoin off-flavors.

Selenomonas spp. Cause the appearance of acetic, lactic and propionic acids in final beer.

Zymophilus contribute with the acetic and propionic acids.

Clostridium provides butyric, caproic, propionic, and valeric acids.

Brevibacillus spp. Results in no off-flavors.

Indicator microorganisms

Certain microorganisms appear as indicators of inappropriate cleaning or production faults. Some examples are [20]:

Acetobacter spp.,
Acinetobacter calcoaceticus,
Gluconobacter oxydans,
P. agglomerans,
Klebsiella spp. And aerobic wild yeasts belong to this category as well [6].

Latent microorganisms

They can occasionally be detected in beer meaning they withstand all unit operations during production process. Frequently these microorganisms can be found in soil and water thus they originate from the water or construction works in the brewery. In case they are detected often they are treated as a sign of unsuitable hygiene. Spore-forming bacteria, enterobacteria, micrococci and film-forming yeast species are comon microbes related to brewing process [6].

2.2 Dispensing Systems Contamination

As can be seen from the previous sections, microorganisms can end up in beer during production. However, it is more likely that microorganism end up in a glass via dispensing equipment than due to the brewing or packaging process error. There is possibility of beer contamination due to microbiological contamination during the process as well as due to inappropriate keg hygiene but the most significant risk is due to bad hygiene of dispensing system i.e. inappropriate sanitation.

The presence of oxygen and warmer environment in the dispensing system contribute to possibility of contamination. Oxygen in the dispensing tap and at the keg tapping head, the pipe lines that may be oxygen permeable [21] can provide excellent environment for microbial growth and contaminate the draught beer. Dispensing lines are usually not completely cooled, especially near the tap where a non-cooled area often appears. Such regimes (combination of $> O_2$ and $> T$) are favorable for acetic acid bacteria, fair number of coliforms and aerobic wild yeast [22–26].

Dispensing systems are favorable environment for various bacteria and yeasts [21, 22, 25, 27]:

Acetobacter,
Gluconobacter,
Obesumbacterium,
Lactobacillus (among them *L. brevis*),
Pediococcus,
Zymomonas,
Brettanomyces/Dekkera,
Debaryomyces,
Kloeckera,
Pichia,
Rhodotorula,
Saccharomyces (brewing and wild yeast strains),
Torulopsis.

Acetic acid bacteria isolated from dispensing systems are able to grow in a microaerophilic ambient, in opposite to similar strains adapted for laboratory cultivation [22].

Escherichia coli is a pathogen belonging to coliform bacteria, characteristically it is acid-resistant and commonly related with severe enteric infections [28, 29]. *E. coli* is contagious at a low dosage, due to its tolerance toward gastric juice and ends up in the intestinal tract of the consumer [29]. The possibility for contamination is much higher in pubs and restaurants that serve food and beer.

Kegged beer is microbiologically safe upon leaving the brewery [6, 25, 26]. However, kegs can often get contaminated after being connected to a dispensing system. Even fresh kegged beer can conduct a contamination [21–23, 25], since microbes can successfully bypass the one-way valves.

Dispensing systems in bars and pubs are exposed to microorganisms through the open tap and during keg replacement. Microorganisms can be detected in the whole dispensing system, especially in the dead space areas (keg tapping heads, dispensing taps, manifolds). Namely, the largest surface is the dispensing line, in which microorganisms build up and adhere creating a biofilm [21].

2.3 Dispensing System Sanitation

Microbial biofilms in brewing represent a serious problem. In breweries, biofilm tends to consist of proliferating communities consisting from bacteria, yeast, and fungal genera creating a coating of mucosal buildup that forms a very sticky shield. As biofilm continues to grow and mature in the presence of nutrients, it becomes harder to control and eliminate it and there is possibility of inadequate elimination during the pasteurization. Biofilms cause off-flavors and can be detected in pasteurizers and conveyor systems.

Isolates of

L. brevis,
E. agglomerans and
Acetobacter sp.

can latch to smooth surface materials used in breweries (such as Teflon or stainless steel) [30]. Problematic areas, places where biofilms proliferate, are usually near the fillers, can and bottle warmers [31]. However, they can be found on side rails, wear strips, internal and external areas of conveyor carriages, drip pans, struts linking the chains and on the bottom of and between chain links.

Common microbes determined in biofilms associated usually belong to bacteria of the genera [31]:

Pseudomonas,
Enterobacter,
Klebsiella,
Alcaligenes,
Flavobacterium,
Lactobacillus,
Bacillus and.
Arthrobacter.

Yeast and molds can also be found and they usually represent genera [31]:

Saccharomyces,
Candida,
Rhodotorula,
Trichosporon,
Cladosporium,
Penicillium,
Geotrichum,
Trichoderma,
Mucor,
Hormonconis,
Aureobasidium and.
Paecilomyces.

Dispensing system lines made of polyvinyl chloride (PVC), polythene and nylon also sustain biofilm formation [22]. According to [21], contamination of dispensing systems relates to the formation of organic film adsorbed onto the PVC pipe. The organic material creating a biofilm is made up of mostly polysaccharides or glycoproteins that originate from wort or substances from yeast cell wall. Electrostatic interactions attract microbial colonies to the pipe surface but they cannot adhere on the conditioning film due to the close range charge repulsion. Yeasts have extending surface fimbriae, with which they latch onto the conditioning film. This matrix has the ability to harden and adopt a harder structure, resulting in fairly difficult elimination of biofilm [21]. Cleaning can kill the cells in the biofilm but the remaining settlements enables recolonisation upon refreshing the dispending system with new cells [2].

Control strategies

Microbial spoilage of beer can be prevented by reducing the origins of contamination. Since beer production process is not completely aseptic, contaminations are possible. However, they can be reduced to a minimum by applying fast methods for determination of the number of contaminating organisms [32]. The most important being CIP (cleaning in place) applied according to the master plan based on the risk assessment, as well as cleaning of whole filling area.

Common hygienic control strategies that can reduce microbial contamination regarding food and beverage industry involve [2]:

- pH adjustment, addition of antimicrobial compounds, reduction of water activity,
- increased osmotic pressure
- physical removal of microbial biomass (filtration, elevated temperatures (cooking, pasteurization, low temperature storage)
- sanitary design of equipment used for production (suitable materials and elimination or minimization of dead spaces, eliminating rough surfaces)
- physical separation of spaces and operations (separation of raw materials, people, air and utensils)
- effective, proper and on-time cleaning and disinfection of equipment and facilities.

Cleaning methods and procedures

Since draught beer system differ from pub to pub, it is crucial to adjust the procedure to each system. Nevertheless, there are general principles that can be applied to every system. In order to effectively clean the draught system, cleaning solutions have to be able to enter and reach every bit of beer line, connectors and hardware [3].

Certain items (couplers and faucets) can be hand-cleaned, but most of the system must be washed out with cleaning agents. Currently, there are two efficient cleaning procedures for beer lines:

- recirculation using an electric pump—recommended for all systems.
- static or pressurized canister purification—an alternative to recirculation, it is less efficient and effective.

Recommendations in a nutshell

Draught line cleaning should be conducted every 14 days and the following procedure should be applied [3]:

- remove the beer from lines by using warm water.
- wash out lines with 2–3% caustic solution (worn out lines). Acid- or silicate-based cleaners are also acceptable. The cleaning solution should be at 25–45 °C.
- the caustic solution should circulate through the lines for at least 15 min at a steady flow rate.
- all faucets and hand-clean couplers should washed by hand after disassembly.
- lines should be flushed with cool fresh water until pH matches that of your tap water and no visible debris is being flushed from the lines.

More serious acid cleaning should be conducted every 3 months using the following procedure:

- rinse beer or caustic cleaner (acid cleaner is addition to caustic cleaner) from lines with warm water.
- wash out lines using acid cleaner (temperatures as above)
- circulate the acid solution through the lines for 15 min as stated above and then cleanse lines with fresh water until the pH matches that of tap water and no visible debris is being carried from the lines.

Every 6 months hardware should be cleaned:

- disassemble, service, and hand-clean all fob devices (beer savers, foam detectors).
- disassemble, service, and hand-clean all couplers.

Along with sanitation with caustic solution there is also mechanical cleaning of pipes every 14 days with small sponge ball prior to rinse with clean water.

3 Conclusions

Draught beer is readily welcomed by the consumers, but often unattended taps and dispensing systems, or inappropriate keg storage temperatures can discourage the consumer from ordering the next glass.

Maintaining a good quality draught beer product is very hard, especially in bars and pubs where certain beer styles are not ordered on an everyday occasion. However, following a good hygienic practice recommendations and using proper cleaning agents can easily reduce the possibility of microbial contamination. Of course, with decrease in microbial safety, all other beer properties and quality can be deteriorated.

A continuous and dedicated sanitation and control by applying the required cleaning practice, should result in a maintained quality of draught beer.

References

1. Devolli, A., Kodra, M., Shahinasi, E., Feta, D., Dara, F.: Hygienic control in beer bottling and dispensing system. J. Hygienic Eng. Des. **15**, 5–11 (2016)
2. Storgårds, E.: Process hygiene control in beer production and dispensing. Dissertation, Faculty of Agriculture and Forestry of the University of Helsinki (2000)
3. Association, B.: Draught Beer Quality Manual/Prepared by the Technical Committee of the Brewers Association, 4th edn. Brewers Publications, Boulder, Colorado, USA (2019)
4. Ingledew, W.M.: Effect of bacterial contamination on beer. A review. J. Am. Soc. Brew. Chem. **37**, 145–150 (1979)
5. Haikara, A.: Beer spoilage organisms. Occurrence and detection with particular reference to a new genus Pectinatus. Technical Research Centre of Finland, Publications (1984)
6. Back, W.: Farbatlas und Handbuch der Getränkebiologie. Teil 1. Kultivierung/Methoden, Brauerei, Winzerei. Verlag Hans Carl, Nürnberg, Germany (1994)
7. Dowhanick, T.M.: Advances in yeast and contaminant determination: The future of the so called 'rapid' methods. Proc. J. De Clerck Chair VI, Leuven 11– 15 September 1994.
8. Donhauser, S., Jacob, F.: Warum zählt Bier zu den reinsten Lebensmitteln? Brauwelt **128**, 1452–1461 (1988)
9. Seidel-Rüfer, H.: *Pectinatus* und andere morphologisch ähnliche gramnegative, anaerobe Stäbchen aus dem Brauereibereich. Monatsschrift Brauwissenschaft 101–105 (1990)
10. Funahashi, W., Suzuki, K., Ohtake, Y., Yamashita, H.: Two novel beer spoilage *Lactobacillus* species isolated from breweries. J. Am. Soc. Brew. Chem. 64–69 (1988)
11. Nakakita, Y., Takahashi, T., Sugiyama, H., Shigyo, T., Shinotsuka, K.: Isolation of novel beer-spoilage bacteria from the brewery environment. J. Am. Soc. Brew. Chem. 114–117 (1998)
12. Haikara, A.: The genera Pectinatus and Megasphaera. In: Balows, A., Truper, H.G., Dworkin, M., Harder, W., Schleifer, K.H. (eds.) The Prokaryotes, A Handbook on the Biology of Bacteria, 2nd edn., pp. 1993–2004. Springer-Verlag, New York (1992)
13. Schleifer, K.H., Leuteritz, M., Weiss, N., Ludwig, W., Kirchhof, G., SeidelRüfer, H.: Taxonomic study of anaerobic, gram-negative, rod-shaped bacteria from breweries: emended description of *Pectinatus cerevisiiphilus* and description of *Pectinatus frisingensis* sp. nov., *Selenomonas lacticifex* sp. nov., *Zymophilus raffinosivorans* gen. nov., sp. nov and *Zymophilus paucivorans* sp. nov. Int. J. Syst. Bacteriol. **40**, 9–27 (1990)
14. Seidel-Rüfer, H.: Pectinatus und andere morphologisch ähnliche gramnegative, anaerobe Stäbchen aus dem Brauereibereich. Monatsschrift Brauwissenschaft **43**, 101–105 (1990)

15. Back, W.: Bierschädliche Bakterien. Taxonomie der bierschädlichen Bakterien. Grampositive Arten. Monatsschrift Brauwissenschaft **34**, 267–276 (1981)
16. Priest, F.G.: Gram-positive brewery bacteria. In: Priest, F.G., Campbell, I. (eds.) Brewing Microbiology. 2nd edn, pp. 127–161. Chapman & Hall, London (1996)
17. Van Vuuren, H.J.J.: Gram-negative spoilage bacteria. In: Priest, F.G., Campbell, I. (eds.) Brewing Microbiology, 2nd edn, pp. 163–191. Chapman & Hall, London (1996)
18. Stenius, V., Majamaa, E., Haikara, A., Henriksson, E., Virtanen, H.: Beer off-flavours originating from anaerobic spore-forming bacteria in brewery adjuncts. In: Proceedings of the 23rd Congress European Brewery Convention, pp. 529–536. Lisbon, IRL Press, Oxford (1991)
19. Hawthorne, D.B., Shaw, R.D., Davine, D.F., Kavanagh, T.E., Clarke, B.J.: Butyric acid off-flavours in beer: origins and control. J. Am. Soc. Brew. Chem. **49**, 4–8 (1991)
20. Gavini, F., Mergaert, J., Beji, A., Mielcarek, C., Izard, D., Kersters, K., De Ley, J.: Transfer of *Enterobacter agglomerans* (Beijerinck 1888) Ewing and Fife 1972 to *Pantoea* gen. nov. as *Pantoea agglomerans* comb. nov. and description of *Pantoea dispersa* sp. nov. Int. J. Syst. Bacteriol. **39**, 337–345 (1989)
21. Casson, D.: Microbiological problems of beer dispense. Brewer **71**, 417–421 (1985)
22. Harper, D.R.: Microbial contamination of draught beer in public houses. Process Biochem. **16**, 2–7, 21 (1981)
23. Ilberg, V., Schwill-Miedaner, A., Sommer, K.: Mikrobiologische Probleme an Schankanlagen. Brauind **80**, 992–995 (1995)
24. Schwill-Miedaner, A., Ilberg, V., Sommer, K.: Der Hygienestatus von Schankanlagen. Brauwelt **136**, 395–400 (1996)
25. Storgårds, E.: Microbiological quality of draught beer—is there reason for concern? In: European Brewery Convention Symposium Draught Beer, Packaging & Dispense, Edinburgh, September 1996. Verlag Hans Carl, Nürnberg, EBC Monograph XXV, pp. 92–103 (1997)
26. Taschan, H.: Mikrobiologische Untersuchung von Bieren aus Schankanlagen in der Gastronomie. Brauwelt **136**, 1014–1016 (1996)
27. Thomas, K., Whitham, H.: Improvement in beer line technology. In: European Brewery Convention Symposium Draught Beer, Packaging & Dispense, Edinburgh, September 1996. Verlag Hans Carl, Nürnberg, EBC Monograph XXV, pp. 124–137 (1997)
28. Semanchek, J.J., Golden, D.A.: Survival of *Escherichia coli* O157:H7 during fermentation of apple cider. J. Food Prot. **59**, 1256–1259 (1996)
29. Park, S., Worobo, R.W., Durst, R.A.: *Escherichia coli* O157:H7 as an emerging foodborne pathogen: a literature review. Crit. Rev. Food Sci. Nutr. **39**, 481–502 (1999)
30. Czechowski, M.H., Banner, M.: Control of biofilms in breweries through cleaning and sanitizing. Mast. Brew. Ass. Amer. Tech. Quart. **29**, 86–88 (1992)
31. Banner, M.J.: Perspectives on conveyor track treatment. Mast. Brew. Ass. Amer. Tech. Quart. **31**, 142–148 (1994)
32. Hammond, J., Brennan, M., Price, A.: The control of microbial spoilage of beer. In: Proceedings of the 25th Conv. Inst. Brew.—Asia Pacific Section, Perth 22–27 March, pp. 63–69 (1998)

Innovation Management as an Important Segment of the Maintenance and Business Development in IT Companies

Dominika Crnjac Milić(✉)

Faculty of Electrical Engineering, Computer Science and Information Technology in Osijek, Josip Juraj Strossmayer University of Osijek, Kneza Trpimira 2B, 31000 Osijek, Croatia
dominika.crnjac@ferit.hr

Abstract. Innovation is becoming an increasingly common concept seen as a necessity for survival and progress in today's business world. For many organizations, innovation still seems unattainable, and the reason for this is often prejudice, old or bad organizational culture, and the mindset in all structures of the organization, from management to employees. The concept of innovation is often defined as something completely new, unprecedented and associated with the notion of an innovative product or service that a company offers to the market, but the concept has a broader meaning. Innovation has become particularly necessary due to rapid technological development, shortening of the product life-cycle, globalization of markets and greater competitiveness. The aim of this paper is to provide an overview of recent research results published in peer-reviewed publications related to the development of innovation management in the IT sector from three aspects: mindset, process and outcome.

Keywords: Innovation management · IT companies · Innovative mindset

1 Introduction

Today, expected management outcomes include continuous innovation of products, processes, marketing models and approaches, the information supply chain, and business model innovation in general. The continuous and accurate traceability of companies' activities is of increasing importance for their survival and progress. The development of technologies and the more widespread application of artificial intelligence, machine learning, the Internet of Things, virtual reality, robotics, automation, biotechnology, etc., also have a significant impact on the way companies operate employee and management development. According to [1], the mindset and attitude of employees towards innovation and the development of an organizational culture that supports change in the business environment are also crucial. When recruiting employees, more and more attention is paid to the willingness and ability to solve problems in a different way, outside of their usual framework. A breadth of knowledge and an interdisciplinary approach to solving problems and developing innovative solutions are expected, as well as a willingness to

D. Blažević et al. (Eds.): OTO 2022, LNNS 592, pp. 186–197, 2023.
https://doi.org/10.1007/978-3-031-21429-5_18

take risks and accept challenges. According to [2], developing innovation strategies is a major challenge for managers because innovation as such is often unknown in terms of the outcome that is achieved in conjunction with existing business practices and whether it is balanced across the business portfolio. There are tools available to help them compare their company's innovations with those of other companies in order to formulate an innovation strategy while minimizing the possibility of bad decisions. Regardless, the risk associated with the implementation of innovations in business may influence the manager's attitude that the innovations are for a different company or business unit than the one in which they operate. As [3] points out, innovation management is a struggle through several phases over a long period of time, and it is important to be aware that the need to start over may arise at any time.

Managers'/leaders' attitudes toward innovation can vary, some of them inclined to change, others more inclined to maintain the status quo. According to [4], innovation is a process that requires creativity and a certain amount of freedom to develop new ideas, but also continuous learning. It is often not easy to discover innovative employees and give them the opportunity to realize their ideas. On the one hand, the reason for this may be that they do not like to be exposed, on the other hand, they may be afraid that their idea will be rejected or that someone will copy it. As indicated in [5], open communication and working in teams can greatly contribute to employees to dare to share their ideas with colleagues and superiors. They need to be made aware that criticism is important and that they should have a positive attitude toward it. It should also be emphasized to them the importance of accepting feedback from colleagues and superiors, which they can use to improve their proposals. In the IT sector, which is pointed out in [3], this is often in contradiction to the presence of introverted employees who tend to work independently, and often lack problem-solving skills, critical thinking, innovation and creativity.

IT companies often use agile methodology in their daily work, where employees have a lot of tasks and final sprints every day. Innovative but quick solutions are expected from agile teams. According to [6], the above is not conducive to the development of innovations, which requires time and not a stressful environment. Changing the way of work and the working environment can have a positive impact on the creativity of employees and the generation of innovations. It seems interesting to point out one of the conclusions of the research conducted by the Boston Consulting Group (BCG) in 2020 during the pandemic caused by the COVID 19 virus, which states: "This crisis has provided a unique opportunity to reinvent the workplace. Things that might once have seemed impossible have proved surprisingly workable. With collaborative productivity essential to innovation, the changes will enable companies to become more competitive. And given the employee desires for flexibility, the changes will also allow companies to recruit and retain the best talent" [7]. Also according to [8] "the development of a new product provides more benefits to the company, such as competitive advantages, positive change in the strategic direction, return on investment and profit, improved image, strengthened marketing/brand, attraction of good personnel, development and growth of the company, etc.

Various methods and processes are already used for encouraging creativity and innovation of employees, for example Methods of design thinking, Lean startup, design sprint, Reward programme, Brainstorming, an Open Innovation (OI) Approach, but as indicated by [9] "a good business idea, continuous coordination and integration, application of modern technical and technological knowledge, skills and experience constitute the basis on which innovation is implemented and on which it achieves the desired market effects. An appropriate innovation strategy, which is in compliance with the enterprise's corporate goals, gives the company an opportunity to decide what kind of innovation it wants to develop."

After researching the previously published knowledge on the subject, the paper was written in a hypothesis-driven manner;

- today's business leaders and entrepreneurs must continually educate themselves to boldly drive innovation and innovative ways of working in business and ultimately successfully manage
- in IT companies that should base their business on innovation, the status quo occurs when the way the innovation is managed is not continuously considered
- changing the way people work and the workplace can have a stimulating effect on employees in terms of generating innovative ideas

The paper provides an overview of research results published in primary, secondary and tertiary peer-reviewed sources. The method of analysis, synthesis, induction and deduction of published facts was applied. After the introduction to the topic, the second part of the paper points out the importance of the intellectual capital of the company, which is the basis for an innovative approach to work. A model of the division of intellectual capital is described, which ensures easier control and evaluation of parameters, through which employees and managers can get feedback on the value of their company in this segment, identify any shortcomings and react in time to improve the situation. It then describes some of the methods that companies use to try to promote an innovative approach to solving problems and developing new products, services, processes, ways of working, etc. The third part of the paper highlights the specificities and requirements that managers face today, as well as the necessary skills, knowledge and ways of working that they should have in order to develop innovation and creative solutions together with their employees.

2 An Approach to Innovation in IT Companies

Knowledge leads to innovation. It ensures a competitive advantage, so that by improving the knowledge and skills of their employees, companies can develop their competencies that allow them to have a more competitive position in the market. As stated in [10]: "Companies in the high technology sector are in an unfavorable situation as global manufacturers, under pressure from new, cheaper or technologically superior competitors, seek to penetrate all available markets".

2.1 Importance of Intellectual Capital

In [11] it was pointed out that many companies attach importance to intangible and personal intellectual capital, which is the accumulation of knowledge and skills of employees, their mutual relations, reputation in the market, etc. According to [12], the term intellectual capital does not refer to all the knowledge in the organization, but only that which can be converted into value and enables the achievement of competitive advantages. Figure 1 made according to [11] shows the model of intellectual capital, which belongs to the intangible assets of the company and is a part of the market value. According to it, intellectual capital is divided into three basic parts (human capital, structural capital and consumer capital). Human capital includes knowledge, skills, abilities and experience of employees who use them by participating in the company's business processes, but take them with them when they leave the company. Structural capital is created by the activity of human capital, and includes the organizational structure, business processes, habits, routines, databases, systems, as well as patents, licenses, etc. Consumer capital is the company's ability to ensure business progress in synergy with business partners, customers or service users. It refers to the quality of their relationships, and also has other names, for example in [13] it is called relational capital, and in [14] clients' capital.

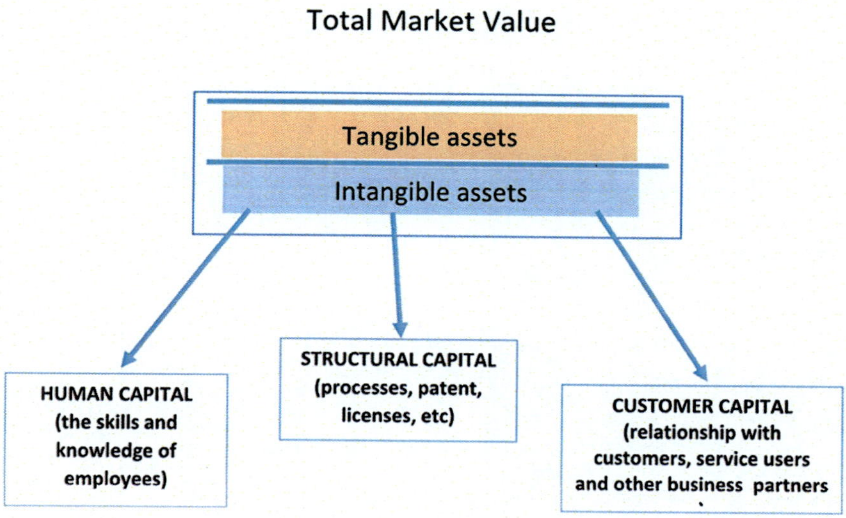

Fig. 1. The intellectual capital model

As pointed out in [15], all segments can be measured, thus ensuring their easier controlling and management. They also give future investors a better picture of the company's market value. Modern companies, especially those in the IT sector, create added value through the interaction of all the previously mentioned forms of intellectual capital. Managers and employees can be more certain of whether they are on the right track in terms of how they work, whether they have achieved a professional degree in a particular field, by having an insight into the situation related to the individual mentioned parameters. Moreover, they can work towards increasing this part of the company's value. As indicated by [16], a high level of knowledge in certain areas of activity can give individuals an incentive to continue working, openness to new solutions, easier communication, and confidence in dealing with colleagues, partners, and customers.

2.2 Methods to Achieve Results

One of the stimulating methods for generating innovative solutions is an Open Innovation (OI) Approach through which companies can use external data more efficiently to generate new ideas. Using this method, companies can upgrade their intellectual capital in order to increase their business value. By applying this approach, individuals and teams responsible for research and development work in companies will have a different role and will have to change the way they work. Figure 2 made according to [17] shows that the company will use its own knowledge, but also what the market offers, i.e. the knowledge of its partners, which can be shared.

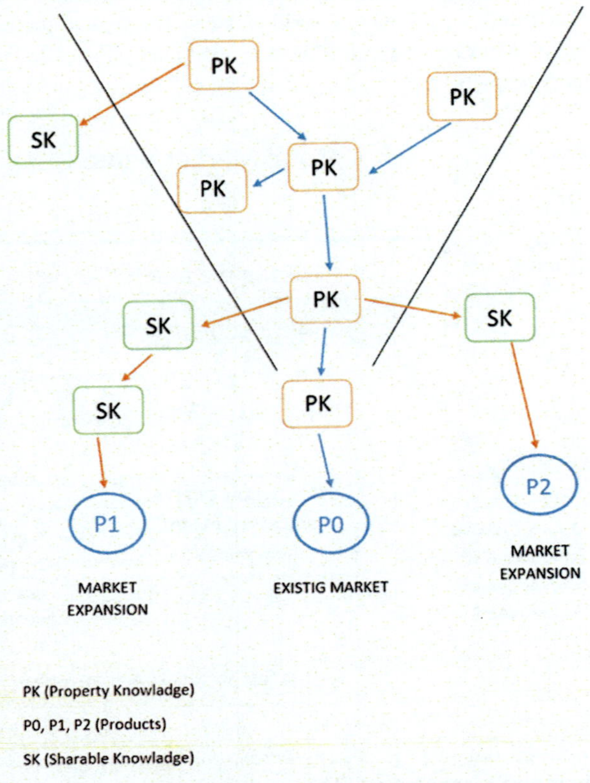

Fig. 2. Open innovation in terms of shareable and proprietary knowledge.

As indicated by [17] "the proprietary knowledge could also join with partners resulting in shareable knowledge and creating a new market. Another possibility is that the combination of proprietary and shareable knowledge will result in new products with various partners." Partnership evaluation is one of the obstacles that can arise along the way, as it is often not easy to integrate the capabilities of the company and its partners. To avoid this problem as much as possible, companies try to seek innovative ideas from their existing partners in the value chain, such as suppliers or customers [18]. As indicated by

[3, 19], there are other methods for idea generation in innovative organizations besides the ones described before. One of them is the development of an agile and innovative culture in such a way that employees are not constantly overwhelmed with work and do not work overtime all the time. The reason for this is that under such working conditions they cannot muster the energy and will to come up with new ideas. It is necessary to give people time to think, but also to reward them for successful work. Design thinking methods are used when the problem comes from the customer, the market or internally and someone needs to be able to design it. It is very important that the problem is defined in a way that the company can solve it, i.e. its employees. They need to understand the challenge and feel that they are capable of solving it. Then they need to brainstorm a solution, prototype it very quickly to test the solution and redesign the idea. At the end, they have to improve the solution. With this method, the company can test the way customers use a certain product and then determine the current and future needs. The main components of the lean startup method are developing ideas, measuring all the data, and then learning and programming. Lean provides the opportunity to approach the customer in a different way before the project begins and can reduce risk and budget for the reason that the Minimum Viable Product is created to begin learning. It is the version of a new product that allows employees to gather facts about customers with the least amount of effort. Design sprint "provide fast forwarding to the future to see the finished product and customer reactions before making any expensive commitments" [20]. It consists of five phases: Understand, Diverge, Converge, Prototype, and Test. In the understanding phase, the focus is on the question, "What is the problem?" in the divergence phase the employees try to find solutions and sketch them, in the convergence phase they decide on a solution and plan the prototype. Then in the prototype phase they design a prototype that can be tested, and in the final phase they test it with real users. The rewards program can vary. According to [3], it must be consistent with the company's culture of innovation and current strategy. Rewarding employees should be based on measurable parameters. For example, they could base it on the number of new idea submissions, the number of submitters and on the quality of ideas connected with the innovation strategy of the organization. Brainstorming offers the opportunity to collect and share ideas in the form of a workshop, where it is important to get participants excited about the topic with a short presentation. They are given explanations that include an up-to-date overview of customers and competitors in relation to the chosen topic and the latest trends in the field. Brainstorming should be combined with innovation task force meetings and an internal open idea competition. In this way, employees get an overview of a complete ecosystem in which their product lives, and silent employees also have another opportunity to participate in innovation initiatives through workshops.

3 Innovation Management

Innovation must be managed. Without proper leadership, innovation will very easily fail to take off. Figure 3, made according to [21] shows the evolution of strategic operations and their management over time. The challenges that managers face are becoming increasingly complex. The continuous progress of technology, the expectations of employees, the quantity of products and services, their quality and the uniqueness expected by the market are constantly changing the ways of management in companies.

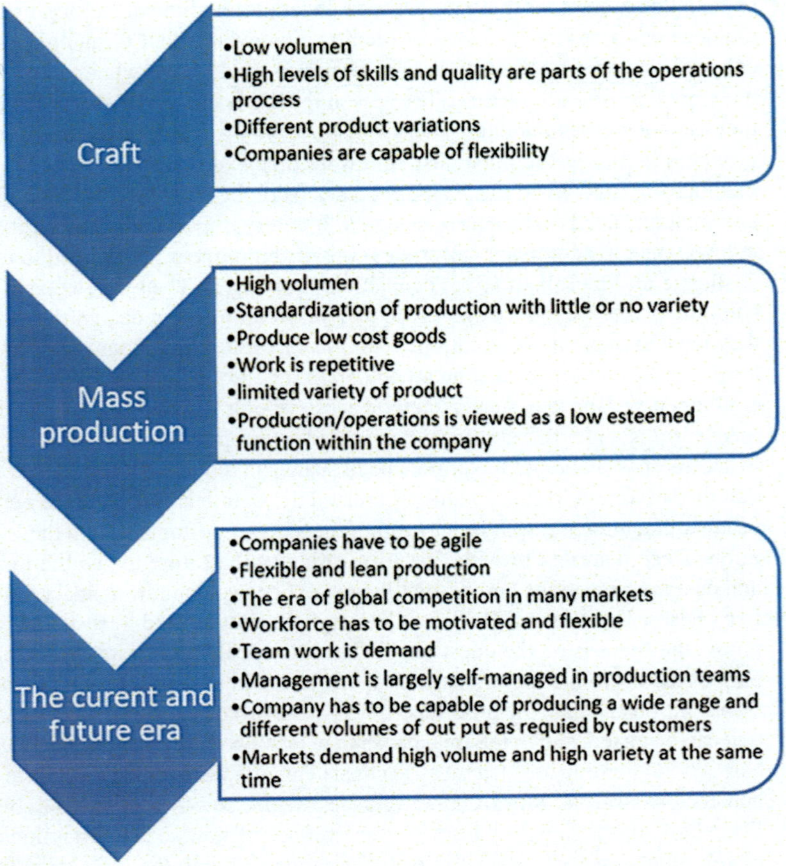

Fig. 3. The transformation from craft to strategic operations

Today's way of doing business requires effective and efficient work, the creation of new products or services in the shortest possible time, at the lowest possible cost and with a quality that satisfies the consumer. For this reason, products/services are often developed through projects. Such a way of working also requires a special management approach.

Project managers must have various knowledge and skills in order to perform their tasks successfully. According to [22] these are: "1. Project Management Fundamentals knowledge and project management software skills (such as the knowledge of using Microsoft Office, email, etc.); 2. Business management skills related to budgeting, finance, procurement, organizational dynamics, team development, performance management, coaching, and motivation; 3. Technical knowledge gained through experience and expertise in the project's focus area (with this knowledge, the manager is more credible, can ask better questions, validate team members' estimates and detailed plans, help solve technical problems, develop better solutions, and provide greater leadership); 4.

Communication skills which include all written communication skills (correspondence, emails, documents), oral communication skills, facilitation skills, presentation skills, and active listening; 5. Leadership skills which includes interpersonal skills, adaptability, flexibility, people management, degree of customer orientation, analytical skills, problem solving skills and so on".

As pointed out in [3], some of the characteristics of innovators in organisations come with time. It is often not easy for managers to deal with them. For example, innovators have a need to satisfy their inner needs, they want to initiate new projects guided by their ideas, they need to break the current status quo, they want to open up new possibilities, visions, and future scenarios, they like to imagine and dream, and they have a strong need to turn their visions into reality.

As stated in [23, 24], it is necessary to develop the organisational culture in such a way that employees are open to suggestions from their colleagues and managers. It is important to pay attention to the fact that employees' creativity, which is the basis for developing innovative products, is not stifled in the sense that they lose interest in work and become demotivated. It is necessary to give them feedback on their proposals and to show them other possibilities if an idea does not fit the situation and is not feasible. It is necessary to encourage research and development because they are often prerequisites for the emergence of innovation in a company. Sometimes innovation can arise spontaneously from the previously acquired knowledge of an individual employee, but according to [10], it most often arises through the planned phases shown in Fig. 4.

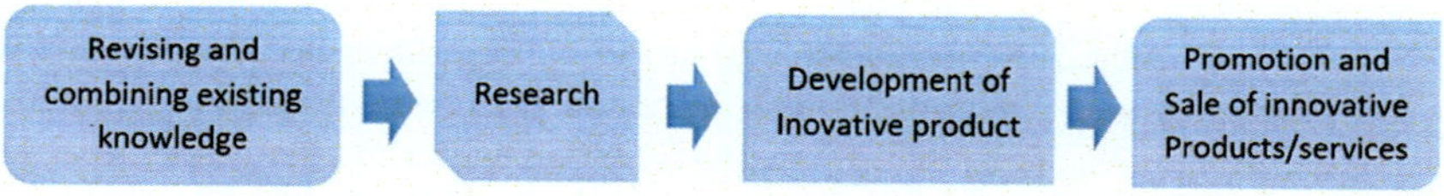

Fig. 4. Development and realization of innovations in the company

Certain innovations are the direct result of previous scientific efforts, so it is preferable to first revise and combine the knowledge acquired so far. If certain necessary knowledge does not exist, then the company starts considering the possibility of investing in research and development. This phase includes creative work in research and development departments, as well as informal or occasional research in other company departments.

Innovations have certain strategic advantages for companies, depending on the mechanism of action with regard to the user. Some of them are shown with Table 1 which is made according to [25, 26].

As pointed out in [27], evaluation criteria should be established at each stage of the product planning and development process. They should be comprehensive and quantitative so that the product can be carefully evaluated at each stage of development. They should be established so that the new idea is evaluated in the context of the market opportunity, competition, marketing mix, financial and production factors. The goal is that it competes with the products/services that exist on the market, i.e. it should have a unique differentiated advantage based on the assessment of competitive products/services that meet the same consumer needs. As indicated in [28, 29], it is

Table 1. Strategic advantages of innovation.

Mechanism	Strategic advantage
The novelty of the product or its maintenance that is offered	Offering something that no one else can offer
New in the process	Offering a way to implement the process in a way that others cannot provide—faster, with lower costs
Complexity	Offering something that others can't do until they pay a license or some other fee
Time advantage	Being the first and having a significant market share related to a new product or service
Platform design	Offering something that provides a platform on which other variations can be built
Rewriting the rules	Offering something that is a completely new product and way of using it, making the old one unnecessary
Reconfiguration parts of the process	Rethinking how parts of the system can work together—building more effective networks, outsourcing, etc.

also important after designing an innovative product/service that through the existing managerial abilities of individuals in the company, marketing strategies and established distribution channels, that product/service achieves a market share. Also, according to [30], it is necessary to take care that it contributes to the financial well-being of the company, it is necessary to monitor the production cost per unit, marketing and distribution costs, the amount of capital required, and the break-even point.

It is also important to take care of existing business processes so that costs related to human resources and equipment do not grow too much. Timely care should be taken of the education of employees and the acquisition of the necessary equipment if necessary in order to meet the quality and quantity demanded by the market. It also seems important to highlight the results of the GEM research (Global entrepreneurship monitoring for 2021), as published in [31], indicates that the Republic of Croatia is building its competitiveness in sectors of medium and high-tech intensity. In terms of the employee entrepreneurial activity (the development of a new product/service or creation of a new business unit for the employer), it has been significantly above the average of European Union countries since 2011, when monitoring of these indicators began. In 2021, 9.4% of employees in the Republic of Croatia had entrepreneurial activity within their company, while the average for the EU countries that participated in the survey is 5.3%. The above should give employers and managers an additional incentive related to the management of innovations in companies, especially in the IT sector, which according to the indicators above is the generator of the competitiveness of the Republic of Croatia on the market. Also according to [32, 33], the state plays a major role in positioning companies in the competitive environment. It can help in the form of a low interest rate for investments and a favorable investment climate. Also important for companies is a well educated and motivated workforce, the training of which is made possible by the well developed education system of the respective country. In addition, the low inflation rate and the rapidly growing domestic market are also important.

4 Conclusion

The focus of this paper is to give an overview of contemporary issues that theorists and scientists are increasingly dealing with in order to improve the operations of companies that must efficiently and effectively deal with the opportunities and challenges brought by globalization. Innovation in products, services, and processes is essential to a company's progress, as well as its survival. It is also increasingly recognized that well-paid jobs based on innovation and knowledge play an important role in promoting the progress of a country. Economic activity is becoming increasingly turbulent worldwide, and competition is fierce in all business sectors. This especially concerns the IT sector, whose market knows no borders, and the speed of change in the products and services they create every day is extremely high. It is recognized that human potential plays a key role and that the term innovation today has a much broader meaning than before. It includes the development and adoption of new types of products, services, production processes and business and organizational models. The promotion of creative and innovative thinking largely depends on the organizational culture of the company.

It can be such that it encourages and supports individuals and teams to be generators of innovation or limits and slows down their action. Policies, procedures and processes that take place in companies also play a big role in creating an enabling environment. Companies that do not provide permanent education and skill building for their employees, that invest insufficiently in technologies and hold back organizational changes are lagging behind in the global race for innovation advantage. They must be aware that investing in product research and development is a necessity in order to achieve economic progress, although there is always a risk that the expected results will not occur. Their success does not only mean prosperity for their employees, but also income growth and improvement of living conditions on a global level. Companies today are multinational, especially those which belong to the IT sector. With their activities, they influence the wider social community, considering that their products are used worldwide, and due to the nature of their work, they employ people in different destinations. For this reason, the job of a manager today is increasingly complex. Managers of all segments of the company should be open to accepting innovative and creative solutions of products, services, processes and ways of working proposed by employees if they can be implemented considering the company's business environment (internal and external). They should encourage an interdisciplinary approach during the generation of these solutions, sharing of information between employees, teamwork, continuous improvement of skills and knowledge of employees and themselves. As a continuation of this conducted research, it should be useful to do a survey to find out from the managers of IT companies which methods they use in order to come up with innovative ideas from employees. Also, how employees react to certain methods and which ones they accept the most. Try to find out if any other methods are applied in innovation management besides those mentioned in the paper.

References

1. Kahn, K.B.: Understanding innovation. Bus. Horiz. **61**(3), 453–460 (2018). https://doi.org/10.1016/j.bushor.2018.01.011. Last accessed 15 June 2022
2. Prange, C., Schlegelmilch, B.B.: Managing innovation dilemmas: the cube solution. Bus. Horiz. **61**(2), 309–322 (2018). https://doi.org/10.1016/j.bushor.2017.11.014. Last accessed 06 June 2022
3. Buljubašić, T.: Developing Innovation: Innovation Management in IT Companies. Walter de Gruyter GmbH & Co KG. Berlin, Germany (2020). ISBN 978-3-11-065306-9
4. Marrone, M., Kießling, M., Kolbe, L.M.: Are we really innovating? An exploratory study on innovation management and service management. In: 2010 IEEE International Conference on Management of Innovation & Technology, pp. 378–383. IEEE (2010). https://doi.org/10.1109/ICMIT.2010.5492719. https://ieeexplore.ieee.org/stamp/stamp.jsp?tp=&arnumber=5492719. Last accessed 21 June 2022
5. Frow, P., Nenonen, S., Payne, A., Storbacka, K.: Managing co-creation design: a strategic approach to innovation. Br. J. Manage. **26**(3), 463–483 (2015). https://doi.org/10.1111/1467-8551.12087. Last accessed 30 June 2022
6. Shalley, C.E., Gilson, L.L.: What leaders need to know: a review of social and contextual factors that can foster or hinder creativity. Leadersh. Quart. **15**(1), 33–53 (2004). https://doi.org/10.1016/j.leaqua.2003.12.004. Last accessed 27 June 2022
7. Dahik, A., Lovich, D., Kreafle, C., et al.: What 12,000 employees have to say about the future of remote work (2020). https://www.bcg.com/publications/2020/valuable-productivity-gains-covid-19. Last accessed 28 June 2022
8. Stošić, B.: Menadžment inovacija—Ekspertni Sistemi, Modeli i Metode, FON, Beograd, Srbija (2007)
9. Janjić, I., Rađenović, T.: Značaj upravljanja inovacijama u savremenim preduzećima. Ekonomika **65**(3), 45–54 (2019). https://scindeks-clanci.ceon.rs/data/pdf/0350-137X/2019/0350-137X1903045J.pdf. Last accessed 25 June 2022
10. Galović, T.: Uvod u inovativnost poduzeća, Sveučilište u Rijeci, Ekonomski fakultet, Rijeka (2016)
11. Stewart, T.A.: The Wealth of Knowledge: Intellectual Capital and the Twenty-first Century Organization, A Currency Book. New York, United States (2001)
12. Jambrek, I., Penić, I.I.: Upravljanje ljudskim potencijalima u poduzećima—ljudski faktor, motivacija zaposlenika kao najbitniji čimbenici uspješnosti poslovanja poduzeća, Zbornik radova Pravnog fakulteta Sveučilišta u Rijeci **29**(2), 1181–1206 (2008)
13. Bontis, N.: There's a Price on Your Head: Managing Intellectual Capital Strategically Business Press. England, London (1996)
14. Edvinsson, L., Malone, M.S.: Intellectual Capital: Realizing Your Company's True Value by Finding Its Hidden Brainpower. Harper Business, New York, United States (1997)
15. Zajednica za unapređenje intelektualnog kapitala (ZUIK) HGK: Intelektualni kapital, 30 godina teorije i prakse u svijetu i Hrvatskoj, Hrvatska gospodarska komora, Zagreb, Croatia (2021)
16. McCloskey, D.N.: Bourgeois Equality: How Ideas, Not Capital or Institutions, Enriched the World. University of Chicago Press, Chicago, United States (2016)
17. Yuana, R., Prasetio, E.A., Syarief, R., Arkeman, Y., Suroso, A.I.: System dynamic and simulation of business model innovation in digital companies: an open innovation approach. J. Open Innov. Technol. Market Complexity **7**(4), 219 (2021)
18. Thompson, N.C., Bonnet, D., Ye, Y.: Why innovation's future isn't (just) open. MIT Sloan Manag. **61**, 55–60 (2020)

19. Nikolaevna, T.V.: Training methods for generating ideas in innovative organization. World Appl. Sci. J. **27**(1), 83–86 (2013)
20. Knapp, J., Zeratsky, J., Kowitz, B.: Sprint: How to Solve Big Problems and Test New Ideas in Just Five Days. Simon and Schuster, New York, United States (2016)
21. Brown, S., Bessant, J., Lamming, R.: Strategis operations management, 3 izdanje. Routledge, Abingdon, New York, United States (2013)
22. Horine, G.: Absolute Beginner's Guide to Project Management. Pearson Education, Indianapolis, United States (2009)
23. Bahtijarević-Šiber, F.: Strateški menadžment ljudskih potencijala. Suvremeni trendovi i izazovi, Školska knjiga Zagreb, Zagreb, Croatia (2014)
24. Mamede, H.S., Santos, V.: The creative information system. In: Dhillon, G., Stahl, B.C., Baskerville, R. (eds.) Information Systems—Creativity and Innovation in Small and Medium Sized Enterprises, pp. 116–121. Springer, Berlin Heidelberg New York, United States (2009)
25. Atkinson, R.D., Ezell, S.J.: Ekonomika inovacija: Utrka za globalnu prednost. Mate d.o.o., Zagreb, Croatia (2014)
26. Mishra, J.K.: Knowledge Management: Complexity, Learning & Sustainable Innovation. Global India Publications, New Delhi, India (2009)
27. Hisrich, R.D., Peters, M.P., Shepherd, D.A.: Poduzetništvo, Mate d.o.o. Zagreb, Croatia (2011)
28. Von Stamm, B.: Managing Innovation, Design and Creativity. John Wiley & Sons, Glasgow, Scotland (2008)
29. Pasovska, S.: Creation of management strategy of a company with a focus on contemporary technology achievements and innovations. University-Industry Links: Coproducing Knowledge, Innovation & Growth (2017)
30. Kirschen, D.S., Strbac, G.: Fundamentals of Power System Economics. John Wiley & Sons, London, England (2018)
31. Singer, S., Šarlija, N., Pfeifer, S., Oberman Peterka, S.: Što čini Hrvatsku (ne) poduzetničkom zemljom? GEM Hrvatska 2019–2020. In: CEPOR–Centar za politiku razvoja malih i srednjih poduzeća, Zagreb (2022)
32. Vurnek, D., Hodak, Z., Bengez, A.: Konkurentnost gospodarstva Republike Hrvatske. Obrazovanje za poduzetništvo-E4E: znanstveno stručni časopis o obrazovanju za poduzetništvo **9**(1), 79–93 (2019)
33. Gordon, E.E.: Future Jobs: Solving the Employment and Skills Crisis: Solving the Employment and Skills Crisis. ABC-CLIO, Santa Barbara, United States (2013)

Improvement and Maintenance of Communication Tools in Micro-enterprises

Katarina Štavlić[1] (✉), Krešimir Lacković[2], and Igor Štavlić[3]

[1] Polytechnic in Požega, 34000 Požega, Croatia
kstavlic@vup.hr
[2] Croatia University North Center, 48000 Koprivnica, Croatia
klackovic@unin.hr
[3] Lifelong Learning Centre, IGOR, 34000 Požega, Croatia

Abstract. Today's rapid and dynamic market changes require continuous investment in improving and maintaining communication tools to sustain and strengthen micro-enterprises competitiveness. Human resources are the bearers of every communication process and must improve existing and develop new competencies and skills. This paper aims to describe the importance of communication tools in micro-enterprises and to emphasise their application in Croatian micro-enterprises. According to the presented comparative results of two empirical studies, it is evident that there are differences in 2020 compared to 2016 in the types of training conducted among employees of micro-enterprises, the level of application of specific communication skills, and the level of business improvement measured by the level of innovation in the surveyed micro-enterprises.

Keywords: Micro enterprises · Communication tools · Human resources · Innovation

1 Introduction

Micro, small and medium enterprises, both in other countries around the world and in the Republic of Croatia, represent the largest share of the total number of enterprises, which is 99.7% in the Republic of Croatia, and employ about 75% of the total number of employees [1]. In the Republic of Croatia, the criteria for classifying entrepreneurs are prescribed by law and harmonised with the European Union's guidelines. According to the Small Business Development Promotion Act, which complies with Commission Recommendation 2003/361/EC of 6 May 2003, represent natural or legal persons that annually employ less than ten workers realise a total annual income in the amount equivalent to EUR 2,000,000.00, or have total assets if they are profit taxpayers, i.e. fixed assets if they are income taxpayers in the amount equal to EU 2,000,000.00 [2]. The importance of micro-enterprises and their operations for the Croatian economy is reflected in the growth of the number of micro-enterprises by 21% in the period 2016–2020, and despite the impact of the pandemic, in the growth of the number of micro-enterprises in 2020 as well.

D. Blažević et al. (Eds.): OTO 2022, LNNS 592, pp. 198–212, 2023.
https://doi.org/10.1007/978-3-031-21429-5_19

Due to the growing number of micro-enterprises, their increasing influence on the national economy and increasingly demanding market conditions, it is necessary to investigate the extent to which such enterprises change their attitude towards employees and work through continuous training and improvement of communication tools and to what extent this contributes the growth of innovativeness of these companies.

Information and communication technology have become an indispensable part of everyday human life, including in the business field. In order for today's micro-enterprises to survive in an increasingly demanding market and in the conditions of local, regional and global competition, companies must become learning organisations with a long-term policy of employee education and training and with an information and communication infrastructure that enables online learning, simple and fast access to information, and cooperation and communication.

Regardless of the size of the company, in today's world of work, there is locational flexibility, i.e. performing activities and communicating with the market is not necessary from a specific location. Communication is an activity that can take place anywhere and at any time, thanks to various communication tools. The main reason and enablers are an intelligent work environment, workplace flexibility and smart devices. The above represents a winning combination for employers, customers, employees and other market stakeholders.

Various communication tools facilitate business and make it more pleasant, efficient and marketable. For remote or mobile communication and interaction, as well as a step towards digital business, appropriate IT and technological infrastructure (smart collaboration methods that include services in the computer cloud), digital literacy and continuous professional training and improvement of knowledge and skills about advanced communication tools and basic business skills, such as sales, presentation, IT, knowledge of foreign languages, etc. Applying the previous knowledge and skills in a smart environment creates conditions for joint work and innovation, making the company more diverse in the business environment. Ultimately, the economy is driven by companies that know how to take advantage of change, technology and challenge. Modern entrepreneurial companies create a constant flow of high-impact products that create value and drive economic growth by creating new methods, technologies and ideas in the global market [3].

It takes a certain amount of time to see the progress and improvement of the business. Enhancements can be monitored at the level of one company through internal performance indicators, at the level of activities through statistical processing of official data, and at the national or world level by monitoring unified indices, hands and various official reports. One of the indexes that monitors the digital economy and society is The Digital Economy and Society Index—DESI. The aforementioned does not exclusively monitor progress in the business environment but monitors overall digital performance and the improvement of European Union countries in terms of digital competitiveness.

Given that human resources are the carriers of communication and other business activities in a company, it is necessary to build and develop the concept of knowledge management in them. It is a systematic and focused process of building, repeating and using the organisation's collective knowledge to achieve its strategic goals. Knowledge management uses both human creativity and intuition, as well as the potential of information technology [4]. For this reason, more and more companies are deciding to invest in formal and informal education of their employees to improve their business competencies and skills and improve operations continuously. Various formal, non-formal and informal forms of education are available on the market for Croatian companies in the form of multiple seminars, courses, study programs, etc. To save time and reduce labour costs, micro-entrepreneurs and their employees most often undergo training and practices related to the performance of business processes and application of new technologies in business to remain competitive on the existing market and prepare for new markets.

The subject of research in this paper is the application of communication skills in micro-enterprises and their impact on business operations measured by the level of innovation in the analysed micro-enterprises. The aim of the work is to analyse the level of improvement in the application of various communication skills of employees in micro-enterprises, the level of increase in training and education of employees, and the level of business improvement measured by the level of innovation in micro-enterprises, in 2020 compared to 2016. The first part of the paper presents a theoretical framework on the importance of human resources in communication processes and maintenance and improvement of business operations, as well as the importance of communication tools for micro-enterprises innovation. The second part of the paper presents a comparative analysis of empirical research.

2 Human Resources as Carriers of Communication Tools and Innovations in Micro-enterprises

2.1 Importance of Human Resources in the Communication Process in Micro-enterprises

Entrepreneurial business and business success results from the interaction between entrepreneurial behavior and internal and external factors of the entrepreneurial environment. The importance of human resources, as an internal factor in improving business performance, has been the subject of numerous studies. Human resources are the bearers of verbal, non-verbal, electronic and other communication in a company, and the business's success depends on their level and quality of communication. Sahin et al. [5] describe human capital as the individual human resources that an entrepreneur possesses, emphasising the level of education, work experience in the profession, management experience and other knowledge and skills of the entrepreneur, such as motivation, leadership styles, focus and ability to control. Bates [6] showed in the results of his research that there is a positive connection between human resources and business longevity and business success. In his study, Hawkins [7] refers to research by Envick, Murphy and others, according to which the internal factors that influence the behaviour of entrepreneurs and ultimately the success of the business are: human capital, social capital, psychological capital and professional capital. Entrepreneurs can impact and control the mentioned internal factors. To connect the influence and meaning of human resources with business performance, it is necessary to measure and evaluate human capital. Hawkins [7] states that not all elements of human capital can be measured. At the same time, some authors believe that human capital can be measured through the connection with financial income [8] and the impact on economic growth [9].

In his work Contemporary Strategy Analysis, Grant [10] indicates the importance of applying strategic analysis in business and measuring the success of using certain types of resources to make the company more competitive and successful. He classified the resources that influence competitiveness and performance into the following categories: financial resources, physical resources, human resources, technological resources and reputation. Human resources include specialised knowledge and skills, communication and interaction abilities, and motivation. In addition to identifying human resources as a factor of success, he proposed key indicators of the success of using certain types of resources, according to which the competitive advantage and success of the company would be measured. The critical indicators for the human resources category are: employees' educational, technical and professional qualifications and the salary ratio to the industry average. As key indicators for the technological resources category, he states the total number and importance of patents, income from sold licenses and the number of employees in research and development as a percentage of the total number of employees. From the research mentioned above, it is evident that employees' communication skills and knowledge and skills are key for the quality business of entrepreneurs. However, business conditions differ in larger than small and micro enterprises.

In micro-enterprises, the organisational and ownership structure is different compared to small, medium and large enterprises. Therefore, in micro-enterprises, the entrepreneurial team or human resources is mainly represented by one person who is usually the owner, entrepreneur and manager who is the decision-maker and who identifies business opportunities and mobilises company resources for the purpose of development and successful business. In micro-enterprises, there are usually no management teams. Still, management teams are comprised of all employees who actively participate in decision-making and implementing business processes that affect the final result of the business. Here, the owner has a primary say, is the responsible person, and is often the only company employee.

Due to the importance of entrepreneurial knowledge and skills of business owners in micro-enterprises, it is necessary to specify this knowledge and skills. Entrepreneurial knowledge and skills have been the research subject for the past 30 years. So De Faoite et al. [11] highlight the following essential entrepreneurial skills:

- the ability to self-assess one's strengths and weaknesses,
- communicating with other people,
- negotiation,
- conflict resolution,
- making decisions,
- planning your own time and energy,
- execution of agreed obligations and
- troubleshooting.

Hisrich and Peters [12] highlight the following skills:

- technical skills: written and oral communication, organisational skills,
- business management skills: planning, decision-making, marketing and accounting skills,
- personal entrepreneurial skills: internal control, risk-taking, innovation, change orientation and visionary leadership.

Micro-entrepreneurs should actively and continuously acquire and improve the aforementioned entrepreneurial skills, regardless of the activity in which they operate, because they are the basis for any business activity. The fact that micro-entrepreneurs are the only bosses in their company and are responsible for the effective management of the company is significant. High-quality management can prevent unwanted effects (employee fluctuations, poor quality of work, etc.) that affect the company's operations. With high-quality management, micro-entrepreneurs can approach each employee as a human being and nurture human values, which is not organizationally possible in larger companies due to reduced communication between leading management and other operational levels.

The characteristics mentioned above, entrepreneurial skills and knowledge are not only crucial for existing entrepreneurs and their success. They are necessary for all individuals who want to be self-employed, manage their life activities independently, and realise their life goals, thus contributing to society as a whole and economic growth and development through their business activity.

In the Republic of Croatia, there is not much research on the communication skills and tools of micro-entrepreneurs. One of the more recent studies by the authors Jelinčić et al. [13] was conducted on a sample of social enterprises, among which micro-enterprises made up 90% of the sample. According to the results of the research, social enterprise employees are aware of insufficiently developed skills and missing knowledge. According to the research results, the following skills and knowledge of employees in social enterprises are most lacking:

1. Financial management skills and knowledge
2. Human resource management
3. Creativity and Innovation
4. Knowledge about the concept of social entrepreneurship
5. Communication and marketing activities
6. Presentation skills
7. Sales skills.

It can be seen that the mentioned research mentions the skills and knowledge related to communication skills and that communication skills are not at the top of the missing knowledge and skills. This paper analyses changes in the application of specific communication knowledge and skills among employees of micro-enterprises in order to determine the connection with the level of innovation, which can lead to a more successful business.

2.2 Communication Tools of Human Resources as a Driver of Innovation in Micro-enterprises

Human work and the application of knowledge, skills and acquired competencies represent the basis of every company's business. Various authors have investigated the importance and impact of human resources on companies' innovative and successful business operations. According to Drucker [4], employees' knowledge and productivity are the company's most valuable assets. Pološki Vokić and Grizelj [14] also wrote about the importance of continuous education and training of employees. They believe that successful organisations are focused on training and improving employees and on their application of knowledge in the improvement, development and growth of business processes. The above can be achieved if the employees have the necessary knowledge and skills to realise the set strategic goals of the business. For this reason, investing in employee education becomes a priority activity, which can result in multiple returns.

We have already mentioned in the previous part of this paper that strengthening employees' communication and other skills improve the business's quality and success. One of the factors of business improvement is innovation, and it can be assumed that improving communication and other skills can increase the level of innovation in a micro-enterprise. The above can be achieved through an innovative communication system within the company. According to the European Commission, Innovation occurs when a company introduces a new or significantly improved product, service, process, marketing strategy or organisational method. Innovation can be developed by the company or was initially developed by another company or society [15].

The connection between innovation and company performance was investigated by Barkham et al. [16], and they observed that companies that undertake innovative activities are not necessarily successful. On the other hand, companies that are the first to offer a new product to the market achieve higher business revenues until their competitors do the same [17]. In addition to product innovations, process innovations are also present in manufacturing companies. Research has shown that companies introducing process innovations achieve higher productivity, improved product quality and lower average production costs [17, 18].

The respondents confirmed the above in one of the interviews [19]. A successful micro-entrepreneur from the fields of agriculture, forestry and fishing stated: "By continuously investing in new technology and equipment, we improve the quality of the production process and ensure the safe placement of finished products, and we have no competition". The existence of a connection between innovation and business success was also pointed out in the results of her research by Darabos, who stated that the results "showed that the entrepreneurship of members results in a higher level of strategic innovation of the company and has a positive impact on the business success of companies in the Republic of Croatia" Daraboš [17]. According to the same research, in addition to entrepreneurship, a higher level of strategic innovation, and thus a positive impact on business performance, is also influenced by the competitiveness and cohesiveness of the management team members. However, there are also perceptions that innovation is not essential for the success of small and medium-sized enterprises. Bill Aulet believes innovations are unnecessary for business success, growth and competitive advantage of small and medium-sized enterprises, tiny enterprises [18]. An increasing number of companies base their operations on innovations and need more significant investments for growth.

Regardless of the results of the previous research, it is evident that human resources and new technologies are associated with the concept of innovation. Today's business communication takes place increasingly through different digital media, and it is necessary to continuously invest in and develop other communication channels and tools to improve business and promote products, services and the company itself more qualitatively and innovatively. It is necessary to strive toward creative and innovative business communication, and certain companies have developed channels of innovative business communication among employees. However, there are also companies where creativity and innovation are not sufficiently motivated, which is considered a waste of time and a more expensive business. A business organisation that is open to innovative communication should have built-in: open channelling of information, clearly defined procedures, and a high level of cooperation among employees, all of which lead to the complexity and dynamic possibilities of innovative communication at all levels (inside and outside the business organisation) [20]. Innovation is one of the key determinants for positioning and defining a successful business organisation in the open market. From an economic and political point of view, innovations are a positive and necessary category for shaping the reputation and creating additional value for the company.

Micro-enterprises in the Republic of Croatia are increasingly aware that their employees are a critical source of innovation. Suppose they do not sufficiently motivate and retain existing employees in their workplaces within the company. In that case, they create potential competition for themselves if employees in their company get an idea for a new product and commercialise it in their own company. This problem is perhaps more common in larger companies with more employees. At the same time, in micro-companies, it is necessary to continuously improve employees' communication tools, motivation and loyalty for continuous growth and development.

3 Research Methodology

From the scientific research available so far and various professional analyses, it is evident that the level of application of communication tools in micro-enterprises and its connection with the improvement of business as measured by the level of innovation have not been sufficiently investigated. Given that this paper focuses on the importance of human resources in the application of communication tools in micro-enterprises, it is necessary to examine the application of their knowledge and skills and the existence of their continuous education. The primary research question was asked: What is the application of communication skills in micro-enterprises and their impact on business as measured by the level of innovation in micro-enterprises?

For the purposes of this paper, some results of empirical research conducted in 2020 and 2016 were used, which were used for previous works by the author Štavlić [19–22], in which the methodology of the conducted research was explained in detail. The basis of the results of this paper are the results of the research carried out in 2020 in the territory of Slavonia and Baranja in the Republic of Croatia, and the comparison was made with the data from the research carried out in 2016 [19]. To determine the level of use of communication skills and innovation in micro-enterprises, an online questionnaire was used for owners of micro-enterprises or decision-makers. The criterion for the classification of micro-enterprises is based on the Act on the Encouragement of the Development of Small Businesses [2] respecting the prescribed criteria regarding the number of employees (less than 10), the achieved annual income (less than or equal to 2 million euros), and that they belong to the agricultural and agricultural activities., processing industry and construction, following the National Classification of Activities [23]. The Lime-Survey software was used to design, implement and generate results. Part of the questions related to the use of communication skills and the presence of continuous education of employees of micro-enterprises were taken from the survey questionnaire from the research Factors of success of micro-enterprises in the Republic of Croatia (2016) [19]. The applied methodology also has its research limitations, which are reflected in the reduced availability of contact data on micro-enterprises, i.e. the absence of an available database with contact data on e-mail addresses or web pages of micro-entrepreneurs. The following limitation was the unfavourable timing of the research implementation due to the resulting pandemic and the reduced interest of micro-entrepreneurs in filling out the questionnaire because they were focused on adapting their business to the newly created conditions. Appropriate statistical and mathematical scientific methods were used to present and interpret the research results.

4 Results and Discussion

The results of this research are based on the responses of surveyed micro-enterprises in five Croatian counties: Brod-Posavina, Osijek-Baranja, Požega-Slavonia, Virovitica-Podravina and Vukovar-Srijem county. Descriptive statistics have determined certain deviations in the distribution of micro-entrepreneurs in 2020 compared to 2016. In 2016, the largest number of responses came from micro-entrepreneurs from Brod-Posavka (24.1%) and Vukovar-Srijem County (23.2%), while in 2020, they came from Brod-Posavka (43.8%) and Pozego-Slavonia county (23.14%). In 2016, the largest share of micro-enterprises was from the processing industry (56.3%), and in 2020 from agriculture, forestry and fishing (36.37%). In 2020, no financial data of micro-entrepreneurs were available, and it was impossible to compare financial data using non-parametric statistical tests.

To determine the existence of the application of specific communication skills and tools among the employees of micro-enterprises, the question of the application of certain knowledge and skills in business was also asked in both surveys. It is also one of the criteria used to determine the application of communication skills in business. The comparative results are shown in Table 1. The results show the rate of change in the arithmetic mean of the application of certain communication skills by the surveyed employees of micro-enterprises.

Table 1 shows the results of applying certain knowledge and skills directly related to the communication skills of employees in micro-enterprises. Based on the respondents' answers in 2016 and 2020 about the application of their knowledge and skills in business and at the workplace, a calculation was made of the rate of change (expressed as a percentage) in 2020 compared to 2016 in the level of use of said knowledge and skill. It can be seen that the employees made the biggest improvement in the use of foreign languages, which is visible through growth of 40.57%. The application of both negotiation (4.48%) and sales skills (1.10%) increased, while there was no change in the application of IT and computer skills. However, the most significant change and negative growth rate are visible in the application of knowledge about modern business improvement methods (−34.64%), and a slightly smaller decrease in the application of knowledge about project management (−16.85%) and analytical skills (−3.21%).

Table 1. Comparative analyses of the application of communication skills by employees of micro-enterprises

Communication skills	Rate of change 2020/2016 (%)
Analytical skills	−3.12
Use of foreign languages	**40.57**
Negotiating skills	4.48
Sales skills	1.10
IT and computer skills	0.00
Project management	−16.85
Knowledge of modern methods for improving business processes (Just in Time, 5S, Kaizen, etc.)	**−34.64**

In the next part of the questionnaire, respondents were asked whether employees in micro-enterprises had participated in any training and courses in the last six months. The comparative results of the answers are shown in Table 2.

According to the presented comparative results of the rate of change regarding the level of education conducted in micro-enterprises, as well as their intentions and plans regarding employee education, it is evident that in the analysed period, there was a decrease in the level of employee education. Namely, there was only a positive rate of change of 15.29% in the attendance of education necessary for performing workplace duties. Despite this growth, in 2020, compared to 2016, there was a decrease in the intention to attend or plan these training in the surveyed companies (−45.65%), and there was an increase in the number of companies that do not plan to conduct the mentioned training (6.38%). The most considerable decrease in the rate of change is visible in education related to communication skills (−69.89%), which indicates a reduction in investment in communication tools in the mentioned companies. The result related to IT courses is also worrying. It can be seen that in addition to the reduced rate of change in attended IT courses in 2020 compared to 2016 in the amount of −54.21%, there was also a decrease in the intention to plan these educations, and the biggest increase was in the number of those who think that they do not pay—they want to educate employees in the field of IT. With the least attended education, communication skills are acquired and perfected, both in a foreign language and through computer literacy, which in the future is the basis for better research and market monitoring and better and more competitive business, based on cooperation and healthy communication with all stakeholders business environment. However, according to the respondents' responses, there are no plans to attend the mentioned training courses in the future, which indicates potential difficulties in digital communication and competition in existing and new markets.

Table 2. Rates of change in the application of communication skills—comparative analyses.

Education	Rate of change 2020/2016 (%)		
	Yes	No, but we plan to	No, nor do we plan to
Managements skills	−41.09	−69.66	111.85
Training necessary for performing job-related tasks	**15.29**	−45.65	6.38
IT courses	−54.21	**−78.79**	**128.33**
Foreign language courses	−44.24	−62.65	42.18
Communication skills	**−69.89**	−58.05	82.67
Other educations	−15.42	−60.40	73.76

In order to determine the level of innovation in micro-enterprises, entrepreneurs were asked whether they had introduced any of the following innovations in their company in the past three years. They were offered answers: a new or significantly improved product, a new or significantly improved service, a new or significantly improved process, a new or significantly improved marketing strategy, a new or improved considerably organisational method, and the number of registered patents. Respondents could choose one of the offered answers, and the results of their answers in the form of the rate of change of the percentage share of responses are in Table 3.

The data in Table 3 shows that in 2020, compared to 2016, there was an increase in innovations in certain areas. Thus, the rate of change for 0 innovations for all presented types of innovations was visibly reduced. The decrease mentioned above in the rate of change increases in other areas because the higher the rate of change in a particular area, the greater the number of innovations and vice versa. The percentage of companies that introduced 1–5 new or improved marketing strategies increased by 91.61%, while the share of companies that introduced 6–15 innovations in the new or significantly enhanced organisational methods increased by 199.76%. At the same time, the highest growth in the rate of change in the number of companies that introduced 16 or more innovations related to organisational methods was recorded in this area, in the amount of 390.51%. On the other hand, there is a visible decrease in the number of companies with 6–15 patents (−18.25%) and the number of companies with 16 or more improved products. From the above, it can be concluded that the largest number of new innovations in micro-enterprises in 2020 compared to 2016 was in the areas of new or improved marketing strategies and organisational methods. The mentioned areas are directly connected with communication tools and processes inside and outside the company, especially with employees' interpersonal communication as carriers of marketing and organisational processes. The minor changes occurred in the innovation of new patents and new products. The obtained results follow the assumptions that innovations are one of the key factors of business success [24, 25]. Based on the results of the empirical research, it can be concluded that micro-enterprises with a higher level of innovation are more successful in business.

Table 3. Rates of change in the number of innovations—comparative analyses.

	Rate of change in the number of innovations in 2020 compared to 2016 (%)			
A type of innovation	0	1–5	6–15	16 and more
A new or significantly improved product	−58.50	83.32	127.09	−18.25
A new or significantly improved service	−53.28	52.36	30.80	186.13
A new or significantly improved process	**−70.69**	36.85	177.96	90.75
A new or significantly improved marketing strategy	−56.91	**91.61**	131.63	0.00
A new or significantly organisational method	−43.80	89.78	**199.76**	**390.51**
Number of registered patents	−0.91	11.48	−18.25	0.00

Communication skills and tools cannot be applied without human labour and employees. According to previous research, human resources are generators of product quality, sales and marketing activities [26]. The improvement and strengthening of business can be generated by strengthening the knowledge and skills of employees, which can be measured by the level of formal and informal education, work experience and specific knowledge and skills [27].

According to the research results presented in this paper, it is evident that there was an increase in the use of certain communication skills by employees in micro-enterprises in 2020 compared to 2019, especially the use of foreign languages and negotiation skills, while there is a visible decrease in the use of project management skills and knowledge about the application of modern methods of improving business processes. It is possible to see the connection of the mentioned results with the results of increasing the number of education and training in the field of IT courses, courses on communication skills, and courses on managerial skills. All of the above points to the increased use of communication tools, which also improves business. And which is visible through the increase in the number of companies that have a greater number of innovations in business.

Today, innovations and innovation activities are considered key factors and drivers of business success and the initiation of overall economic growth. To create innovations through new or improved products, services or processes, it is necessary to invest in the existing business continuously. The visible increase in the number of innovations and the growth in the application of specific communication skills in the analysed companies may also be a consequence of a continuous investment in their operations. In the surveyed companies, the highest investments in purchasing machines, equipment, programs and licenses occurred in the analysed period, and they had an increase in investment in employee training and education [26]. Strengthening employees' knowledge, skills and competencies represents strength and potential for supporting innovation as an essential factor in the success of micro-enterprise operations. According to research results [19], micro-enterprises with a higher level of innovation are more successful. There is a statistically significant positive correlation: between performance and the number of new or improved services, between the financial indicator of profitability and ROE and investment in the improvement of organisational or business processes, and between the performance of companies and continuous investment in research and development, innovation and the application of modern technologies in business processes. The above leads to the conclusion that, regardless of the existing small investments in research and development in micro-enterprises, there is an innovation that ensures successful operations of micro-enterprises. This opens up the possibility of new research on determining the number, type and reasons for innovation in individual micro-enterprises.

To improve the application of communication tools in micro-enterprises, planning and strategic planning of such activities is necessary, as well as human resource management for the purpose of more efficient operations. Strategic planning is still insufficiently present in micro-enterprises [19, 27, 28], which is also evident from the research results, where micro-entrepreneurs broadly do not plan to conduct specific training for their employees in the coming period.

By conducting this research, certain limitations were detected. Research limitations stem from a limited and small territorial and numerical sample, non-availability of digital contact data of micro-entrepreneurs, and unavailability or non-existence of a base of defined competencies, knowledge and skills needed by employees of micro-enterprises in the area of the research. During future research, the established limitations can be taken as a basis when defining the research question, sample, and objectives on communication tools in micro-enterprises. It was undoubtedly helpful to investigate and determine the existence of a statistically significant connection between the level of use of communication tools by employees in micro-enterprises and innovation and business improvement. In addition to the problem of unavailable contact information, additional issues in this research were manifested through a lower rate of completion of the survey questionnaire. The assumption is that the reason for this is the size of the company, i.e. the number of employees (the average number of employees in the surveyed companies is 2.8 employees), and the lack of time and digital skills.

5 Conclusions and Recommendations

The presented results indicate several conclusions about applying communication tools in micro-enterprises in the Republic of Croatia. The application of communication tools in micro-enterprises was measured in this paper using two criteria, the first is the application of communication skills, and the second is the level of education and training of employees in micro-enterprises. In the analysed micro-enterprises, in 2020 compared to 2016, there was increased use of foreign languages and negotiation skills, while analytical skills, project management skills and knowledge of modern business improvement methods were used less. In the same period, there was an increased level of employee education in training necessary for business processes. In contrast, employee education in communication skills decreased by 69.89%. Many micro-enterprises do not plan to educate their employees in the coming period of their operations, especially in IT courses, management skills and communication skills. In the future, those mentioned above can significantly affect the business's competitiveness if employees' non-continuous education is reflected in their weaker communication with internal and external market stakeholders. One of the reasons may be the reduction of business costs through the reduction of investment in employee education. Still, in the long term, it is necessary to continuously invest in the further development of human resources. The results related to the application of communication tools were also reflected in the level of innovation in micro-enterprises, especially in the type of innovation. Namely, there is a visible reduction in the number of patents and innovations related to product improvement. At the same time, there has been an increase in innovations in the field of organisational methods. It is undoubtedly necessary to continuously invest in educating employees and communication infrastructure so that micro-enterprises can use digital and other communication tools to perform in selected markets sustainably. The results indicate that micro-entrepreneurs, regardless of existing resources and market situations, with the aim of competitive and sustainable business, should generate growth and development based on the application of new knowledge and skills of employees.

References

1. CEPOR: www.cepor.hr. http://www.cepor.hr/wp-content/uploads/2015/03/CEPOR-Mala-i-srednja-poduze%C4%87a-u-HR-u-vrijeme-pandemije-COVID-19.pdf (2022). Last accessed 9 July 2022
2. Small Business Development Encouragement Act: National newspapers, Narodne novine no. 29/02, 63/07, 53/12, 56/13, 121/16, Art. 3. Available from: https://www.zakon.hr/z/527/Zakon-o-poticanju-razvoja-malog-gospodarstva (2016). Last accessed 5 July 2022
3. Dvorski, J.: The determinants of innovative communication in the business environment. Ph.D. Dissertation, Faculty of Humanities and Social Sciences, University of Zagreb (2017)
4. Drucker, P.F.: Management Challenges for the 21st Century. HarperCollins Publishers, New York (1999)
5. Sahin, M., Nijkamp, P., Stough, R.: Impact of urban conditions on firm performance of migrant entrepreneurs: a comparative Dutch-US study. Ann. Reg. Sci. **46**, 661–689 (2011). https://doi.org/10.1007/s00168-009-0351-2. http://download.springer.com/static/pdf/774/art%253A10.1007%252Fs00168-009-0351-2.pdf?originUrl=http%3A%2F%2Flink.springer.com%2Farticle%2F10.1007%2Fs00168-009-0351-2&token2=exp=1461150506~acl=%2Fstatic%2Fpdf%2F774%2Fart%25253A10.1007%25252Fs00168-009-0351-2.pdf%3ForiginUrl%3Dhttp%253A%252F%252Flink.springer.com%252Farticle%252F10.1007%252Fs00168-009-0351-2*~hmac=951a5ea699fa4a3a41c5751c3d314803c54aba08910a2c631603077949f22eb9. Last accessed 28 June 2022
6. Bates, T.: Social resources generated by group support networks may not be beneficial to Asian immigrant-owned small business. Soc. Forces **72**(3), 671–689 (1994). http://sf.oxfordjournals.org/content/72/3/671.abstract. Last accessed 28 June 2022
7. Hawkins, W.D.: Exploring success factors of female entrepreneurs in the texas child care market (Order No. 3569855) (2013). Available from ProQuest Dissertations & Theses A&I (1418478828). http://search.proquest.com/docview/1418478828?accountid=44407. Last accessed 6 May 2022
8. Bruns, V., Holland, D.V., Shepherd, D. A., Wiklund, J.: The role of human capital in loan officers decision policies. Entrepreneurship Theory Pract. **32**, 485–506 (2008). https://doi.org/10.1111/j.1540-6520.2008.00237.x/pdf. Last accessed 1 July 2022

9. Gohmann, S.F., Hobbs, B.K., McCrickard, M.: Economic freedom and service industry growth in the United States. Entrepreneurship Theory Pract. **32**, 855–874 (2008). http://papers.ssrn.com/sol3/papers.cfm?abstract_id=1248406. Last accessed 28 June 2022
10. Grant, R.M.: Contemporary Strategy Analysis, 8th edn. Blackwell Publishers, Oxford (2013)
11. De Faoite, D., Henry, C., Johnston, K., Van der Sijde, P.: Entrepreneurs' attitudes to training and support initiatives: evidence from Ireland and The Netherlands. J. Small Bus. Enterp Dev. **11**(4), 440–448 (2004). https://doi.org/10.1108/14626000410567080. Last accessed 1 July 2022
12. Hisrich, R.D., Peters, M.P.: Entrepreneurship. 4, izdanje. Irwin McGraw-Hill, Boston, MA (1998)
13. Jelinčić, D.A., Baturina, D., Franić, S.: Impact of service learning on social entrepreneurship and youth employment in Croatia. Interdisc. Description Complex Syst. **20**(4), 319–335 (2022). https://hrcak.srce.hr/file/405006. Last accessed 9 July 2022
14. Pološki Vokić, N., Grizelj, H.: Obrazovanje i razvoj zaposlenika u hrvatskim organizacijama. Ekonomski pregled **58**(12), 851–880 (2007)
15. European Commission: Innobarometer 2015—The innovation trends at EU enterprises, European Union, ISBN 978-92-79-47769-0 (2015). https://www.google.hr/webhp?sourceid=chrome-instant&ion=1&espv=2&ie=UTF-8#q=INNOBAROMETER+2015+-+THE+INNOVATION+TRENDS+AT+EU:7+ENTERPRISES%2C. Last accessed 1 July 2022
16. Barkham, R., Gudgin, G., Hart, M., Hanevey, E.: The Determinants of Small Firm Growth. An Inter-Regional Study in the United Kingdom 1986–1990. Routledge Taylor & Francis Group, London and New York (1996). https://books.google.hr/books?hl=hr&lr=&id=kA0aFzEK20EC&oi=fnd&pg=PP2&dq=barkham+the+determinantes+of+small+firm+growth&ots=pQKflrVMnb&sig=mFn7ojyA8O7cl8fZRTOYJEJeiVI&redir_esc=y#v=onepage&q=barkham%20the%20determinantes%20of%20small%20firm%20growth&f=false. Last accessed 2 July 2022
17. Daraboš, M.: Evolucija konkurentske prednosti—u potrazi za uspjehom u hiperkonkurentskim uvjetima, Naklada Ljevak d.o.o., Zagreb (2015)
18. Aulet, B.: Disciplinirano poduzetništvo, Mate d.o.o., Zagreb (2015)
19. Štavlić, K.: Success factors of micro enterprises in the Republic of Croatia, Čimbenici uspješnosti mikro poduzeća u Republici Hrvatskoj. Ph.D. Dissertation, Faculty of Economics, Josip Juraj Strossmayer University of Osijek, Croatia (2016)
20. Stavlic, K.: Comparative analysis of the presence of the new and modern technologies in micro enterprises. In: Katalinic, B. (ed.) Proceedings of the 31st DAAAM International Symposium, pp. 0936–0942. Published by DAAAM International, Vienna, Austria (2020). ISBN 978-3-902734-29-7, ISSN 1726-9679. https://doi.org/10.2507/31st.daaam.proceedings.130
21. Štavlić, K.: Analysis of the innovativity of Croatian micro-entrepreneurs. In: Katalinic, B. (ed.) Proceedings of the 7th International Conference "Vallis Aurea" Focus on: Research & Innovation, 24–26 September 2020, Požega, Croatia, pp. 603–611. Polytechnic in Pozega, DAAAM International Vienna, Požega, Vienna (2020). ISSN 1847-8204, ISBN 978-3-902734-26-6
22. Štavlić, K., Potnik Galić, K.: Comparative analysis of the level of education and skills of Croatian micro-entrepreneurs. In: Gomez Chova, L., Lopez Martinez, A., Candel Torres, I. (eds.) INTED2021 Proceedings, pp. 7261–7266. IATED Academy, Valencia (2021) ISBN: 978-84-09-27666-0, ISSN: 2340-1079
23. National Occupational Classification—NKD 2007, NN 58/2007. https://narodne-novine.nn.hr/clanci/sluzbeni/2007_06_58_1870.html. Last accessed 5 May 2022
24. Greenbank, P.: Micro-business start-ups: challenging normative decision making? Mark. Intell. Plann. **4**(4) (2000)
25. Nayak, A., Greenfield, S.: The use of management accounting information in managing micro-businesses. In: Finance and the Small Firm. Routledge, London (1994)

26. Štavlić, K., Štavlić, I., Lacković, K.: The importance of employee education and information and communication tools for maintenance and measuring the performance of micro-enterprises. In: Glavaš, H., Hadzima-Nyarko, M., Karakašić, M., Ademović, N., Avdaković, S. (eds.) Proceedings of 30th International Conference on Organization and Technology of Maintenance (OTO 2021), pp. 390–402 (2021). https://doi.org/10.1007/978-3-030-92851-3_29. ISSN 2367-3389, ISBN 978-3-030-92851-3 (eBook)
27. Kay, J.: Foundations of corporate success: how business strategies add value. Oxford University Press, New York. [Online] (2001) http://www.amazon.co.uk/Foundations-Corporate-Success-Business-Strategies/dp/019828781X. Last accessed 2 July 2022
28. Joyce, P., Seaman, C., Woods, A.: The strategic management styles of small businesses, Chapter 5. In: Blackburn, R., Jennings, P.L. (eds.) Small Firms, Contributions to Economic Regeneration. Paul Chapman Publishing, London (1996). http://resourcelists.ntu.ac.uk/items/BD9AE01F-417D-DE94-F1BC-A8237A2E41C7.html. Last accessed 2 July 2022
29. Čerić, V., Varga, M.: Informacijska tehnologija u poslovanju. Element, Zagreb (2004)
30. Porter, M.E.: Competitive Strategy: Techniques for Analyzing Industries and Competitors. The Free Press, New York (1998)
31. Byers, T.H., Dorf, R.C., Nelson, A.J.: Tehnološko poduzetništvo: Od ideje do tvrtke. Tehnološko Veleučilište u Zagrebu, Zagreb (2015)

Author Index

MISC
Šolić, Tomislav, 118
Štavlić, Igor, 198
Štavlić, Katarina, 198
Švarcmajer, Miljenko, 25
Živković, Andrej, 18

A
Aksoy, Muharrem Hilmi, 169
Ašonja, Aleksandar, 133, 143

B
Bilandžija, Domagoj, 18
Blažević, Damir, 18, 37, 45

C
Cumin, Josip, 118, 150

D
Desnica, Eleonora, 133, 143
Dihovicni, Djordje, 169
Domaćinović, Ana, 176
Duspara, Miroslav, 150

F
Fekete, Kresimir, 57

G
Glavaš, Hrvoje, 96
Grgić, Ivan, 106

H
Habschied, Kristina, 176
Hajkowski, Jakub, 150

I
Ivezic, Dijana, 1

J
Jarčević, Duška, 96
Jovanović, Miloš, 169

K
Kacmarcik, Josip, 160
Karakašić, Mirko, 106
Katinic, Marko, 160
Keleković, Snježana, 176
Keser, Tomislav, 18, 37, 45
Kiss, Imre, 133
Kljajic, Ruzica, 57
Köhler, Mirko, 25
Konjatic, Pejo, 160
Kravchenko, Igor, 143
Kušević, Tomislav, 45
Kuznetsov, Yury, 143

L
Lacković, Krešimir, 198
Lagator, Domagoj, 84
Livada, Caslav, 1
Lukačević, Igor, 84
Lukić, Ivica, 25

M
Malutkiewicz, Karol, 150
Marić, Dejan, 118
Maric, Predrag, 57
Marijić, Jure, 106
Mastanjević, Krešimir, 176

D. Blažević et al. (Eds.): OTO 2022, LNNS 592, pp. 213–214, 2023.
https://doi.org/10.1007/978-3-031-21429-5

Matysiak, Waldemar, 150
Milić, Dominika Crnjac, 186
Milinović, Andrijana, 160

N
Nenadić, Petar, 37
Noskov, Robert, 69
Novaković, Borivoj, 96
Novoselović, Daniel, 118

P
Palinkaš, Ivan, 133
Pastukhov, Alexander, 143
Pavetić, Tanja Mađarević, 176
Pekez, Jasmina, 96
Petrović, Ivica, 69
Pisonic, Filip, 57
Provči, Igor, 69

R
Radojicic, Sara, 160
Radovanović, Ljiljana, 96, 133

S
Stawiński, Grzegorz, 150
Stipanić, Tomislav, 25
Stoić, Antun, 150

V
Vukašinović, Josip, 106